“To generate light,

to seek out in myself not yet seen poetry,

not to separate but to link.”

Gyorgy Kepes 1906 - 2001

NUMBERS AND MEASURE

First Floor Lobby, 2001
18" x 18" Anodized Aluminum
Plates, Laser Etched Black
and Screen Printed

Photo by Skylar Nielsen

ART FOR A HOUSE OF MATHEMATICS

Anna Campbell Bliss

Photo Documentation

Skylar Nielsen

For Gyorgy Kepes (1906-2001)

Publication has been supported by grants from:

Graham Foundation for Advanced Visual Studies
Public Building Program of the Utah Arts Council
Department of Mathematics, University of Utah

Design, Text: Anna Campbell Bliss
Consulting Editor: Ann Poore
Photo Documentation: Skylar Nielsen
Printing by Printers Inc.

Distributed by
American Mathematical Society
201 Charles St., Providence, RI 02904-2294 USA
For book orders call toll-free 1-800-321-4267
visit webpage www.ams.org/bookstore.

Published by Bliss Studio Publication

Library of Congress Control Number:2006900247
ISBN 0975491512

INTRODUCTION

The commission for the Cowles Mathematics Building Expansion at the University of Utah offered a challenge architecturally as well as intellectually: to create an environment for mathematics in a building that consisted of small lobbies on three floors and extensive corridors in an older structure.

In a competition organized by the Utah Arts Council, the Mathematics faculty was exceptional in its range of vision. They asked for art that would identify the building as a "House of Mathematics" and convey its wide-ranging influence. They also sought an environment that was a source of pleasure, intrigue and inspiration. It was a rare and imaginative statement from a building committee.

With long corridors in the older building and limited wall space in the modern, mostly glass addition, a solo work on a single floor would not satisfy those goals. I began to visualize a multifaceted work that extended throughout the building. It would develop structural connections with the arts, sciences and culture while conveying the range and beauty of mathematics. Small units organized in clusters would integrate with the ied architectural spaces and permit developing a subject in a cohesive way.

In the addition I selected lobby walls on three levels to serve as the major focus. They could be seen from various parts of the building and through the exterior glass. Instead of enclosing the content in a standard frame I chose a more dynamic image. Additional groups of plates would extend into the corridors like fingers.

Much of the work was developed with the help of the computer and incorporated mixed media for a more tactile experience. The 18-inch-square aluminum plates were screened and, in some instances, hand painted. The black anodized plates were laser etched and color added by screening.

Many people have been generous with advice on content and technical problems. I would like to thank Professor Jim Carlson, former chairman of the University's Mathematics Department, for his support and encouragement throughout the project. Professors Hugo Rossi and Klaus Schmitt participated, as did other faculty members. Madelyn Garrett, head of the Rare Books Division of Special Collections at J. Willard Marriott Library offered invaluable help in researching its historic manuscripts.

A. C. Bliss

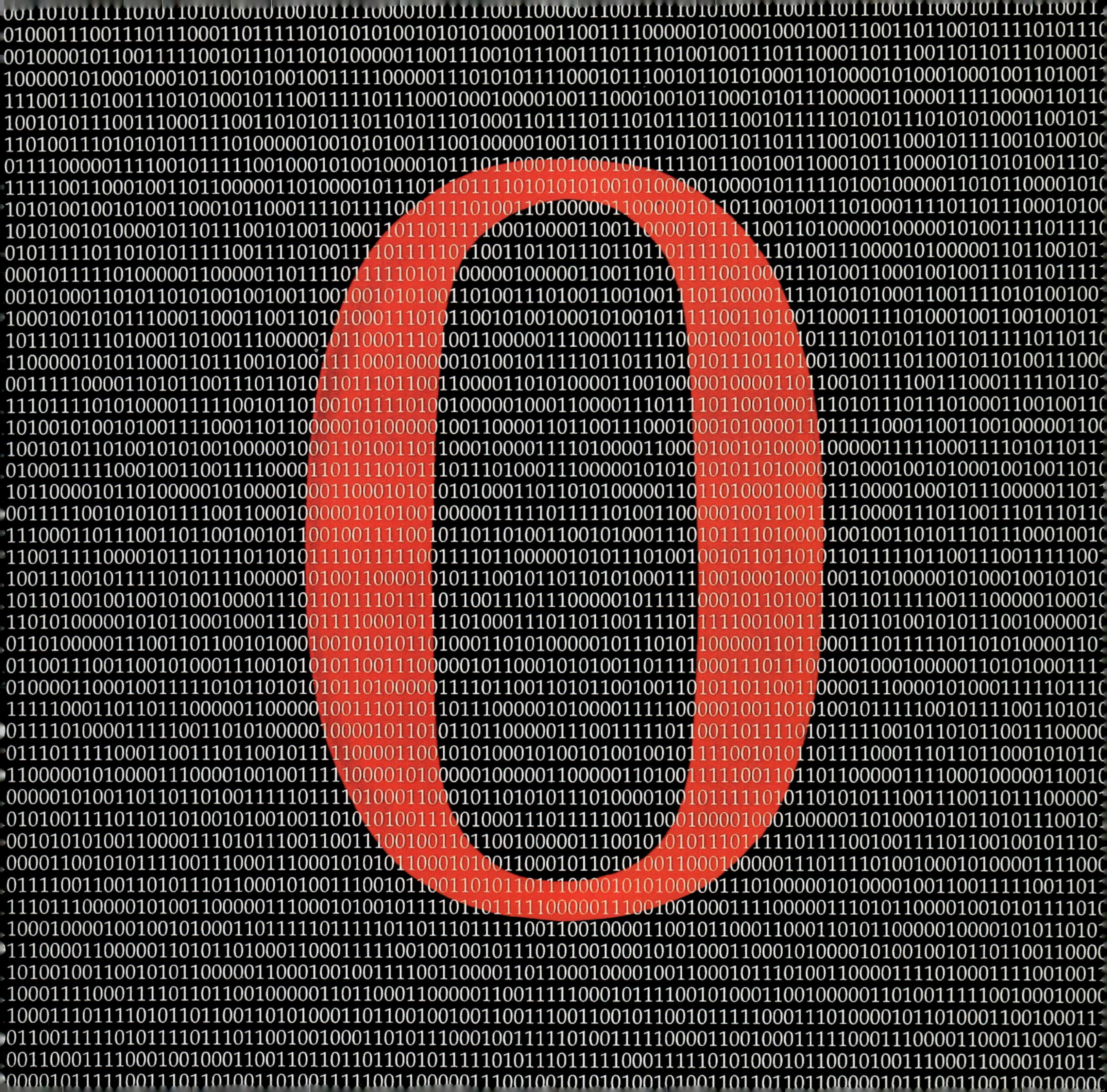

For a complex work of this scope, a strategy was needed to develop the specific art for each area. Various concepts were explored and gradually eliminated. It was not to be a history of mathematics or a literal textbook. The material would be suggestive and a stimulus for the imagination. Personal viewpoints no doubt would influence portions of the art but were balanced by the established criteria. Each plate was to be a work of art but related to its immediate context and part of the general intellectual, mathematical and visual concept.

Numbers were a logical point of entry and a fascinating area of historical research. The centuries-long development of abstract thinking that led to concepts of quantity, sequence, numeration systems and positional rotation could only be inferred here. The numerical sequences of one through nine can be traced to India as early as the third century B.C. But it is not until the concept of Zero emerged as a symbol for nothing and combined with the concept of placement that a modern system of notation developed in the fifth century A.D. From India, Arabs brought it to Europe. Now by its placement Zero could be a value of 10, 100, 1000, etc. using a base of 10. A binary system was adopted for the computer and is shown here as a random bit stream overlaying the Zero. One can appreciate the impact of these developments by considering how difficult it would be today to conduct financial negotiations in Roman numerals.

Plate 13: ZERO
with random bitstream, screen printed and laser etched.

Systems of notation are explored in the group of plates devoted to Mayan, Babylonian and Fibonacci numbers. Traveling in Yucatan one develops a healthy respect for the Mayan culture and frequently observes what appear to be numbers on the sculptural reliefs. Further study reveals a simple but well-developed system of notation. The dot equals one and the bar equals five, drawn here in yellow. A Babylonian system is shown in red. The nail form indicates one and the chevron 10. Including Fibonacci numbers seemed important for artists as well as mathematicians. They, too, have a long history. Developing the numbers over three plates permitted studying the sequences in more detail. It was also a design decision to give emphasis to the subject and the importance of Zero.

The Fibonacci sequence shown in Plates 2, 4 and 9 has many applications in organic growth patterns and in art. One thinks of the beautiful chambered nautilus shell when seeing the equiangular or logarithmic spiral in Plate 16. Although spirals are common in nature, not all exhibit Fibonacci proportions.

Plates

Plates						
1, 2			Modulars 2001 Classic ratio - 1:1.618	Numbers: yellow- Mayan, red- Babylonian, grey- Fibonacci		
3, 4, 5		Choreography by Lucinda Childs, RDT dancers & Alvin Ailey		Numbers: yellow- Mayan, red- Babylonian, grey- Fibonacci	Head studies by Piero Della Francesca, from De Prospectiva Pingendi	
6, 7, 8, 9, 10	Olympic racers: Swatch photo courtesy of Associated Press	Who Counts? Who is Counting? Census 2000	Original jazz composition by Dan Waldis, 2001	Numbers: yellow- Mayan, red- Babylonian, grey- Fibonacci		Aspects of mind. One equals four.
11, 12, 13		Numbers, a poem by Katherine Coles 2001	Islamic calligraphy, measured studies	Zero, with random bit stream		
14, 15, 16			Counting with hand gestures	Renaissance pictorial organization based on the golden number - 1.618	Logarithmic spiral using Fibonacci proportions	
17, 18			Chinese triangle showing Fibonacci numbers with calligraphy by Mishio Zushi	Pascal's triangle showing Fibonacci numbers		

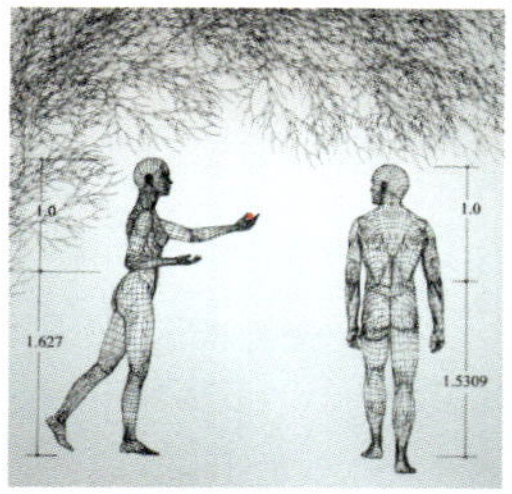

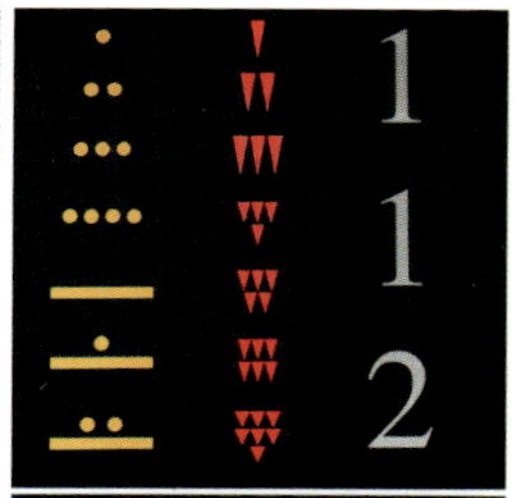

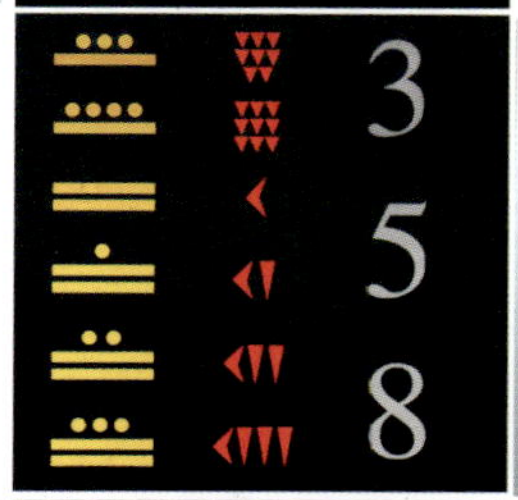

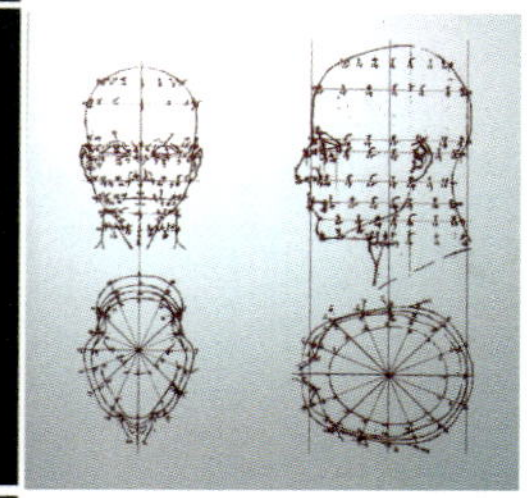

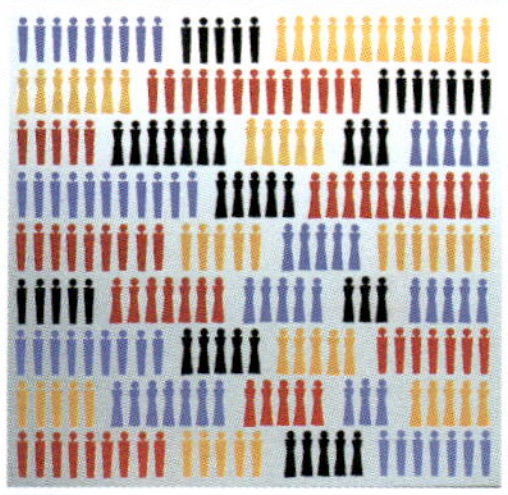

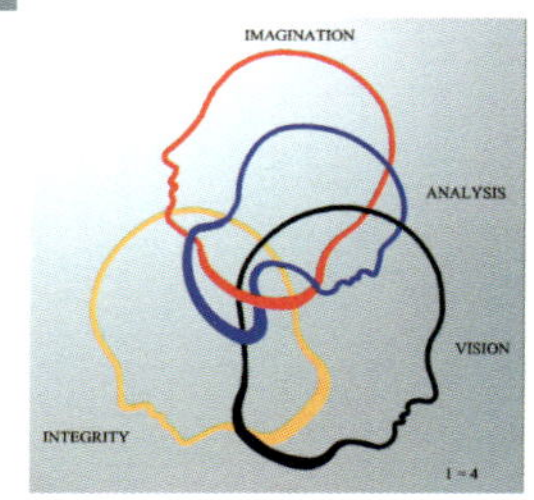

NUMBERS

They won't stick. They gleam like brilliantine.
Perfect parsers, they jostle into essence
then reappear, renewed. A trillion seems
just so many zeroes. Xed-out, they dance,

Uncoupling and recoupling along a line
hoofing infinite movement, can-can's limber
leg and best foot forward, tapping time
until time is up and they're dismembered.

Dead-broke as syntax, clauses so declined
they tick themselves off. It's only beauty–
perfect measures measuring the mind–
mind tries to get around. Pen, brush, or flute.

Equation tooled to figure life. Amount
imagination multiplies. Takes to account.

Katharine Coles

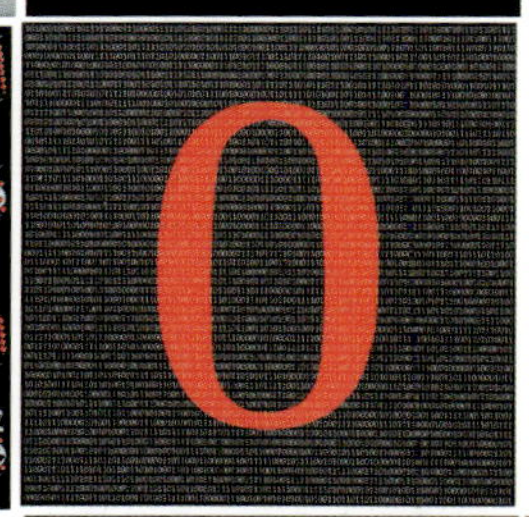

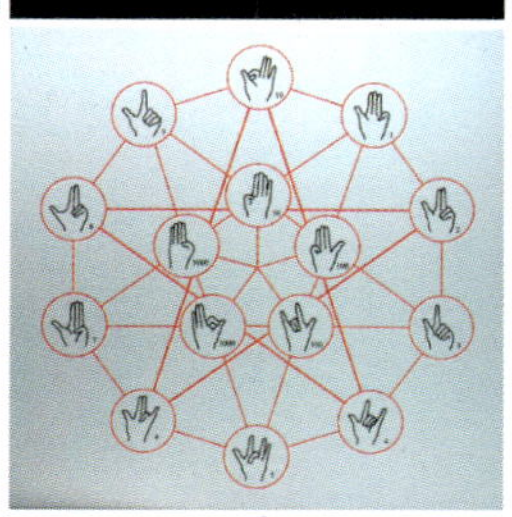

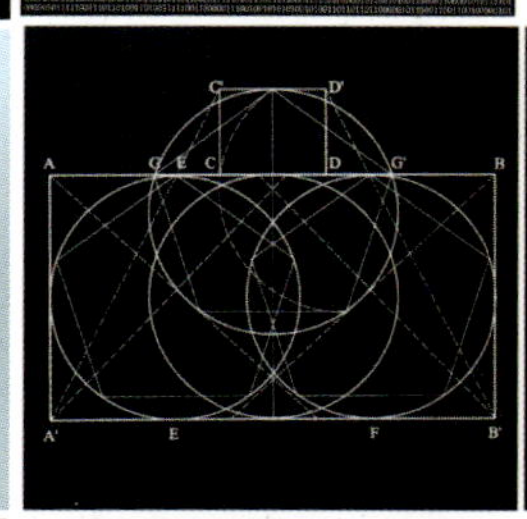

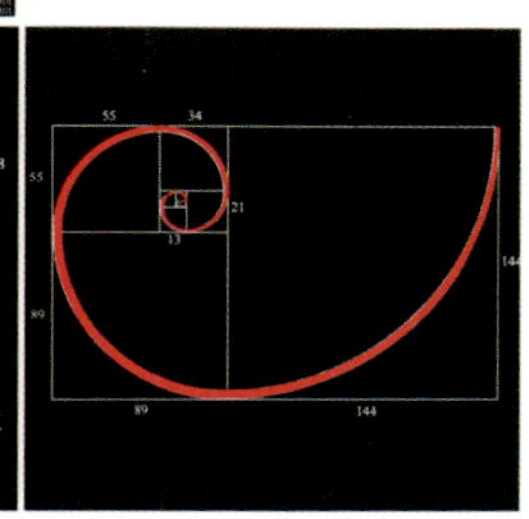

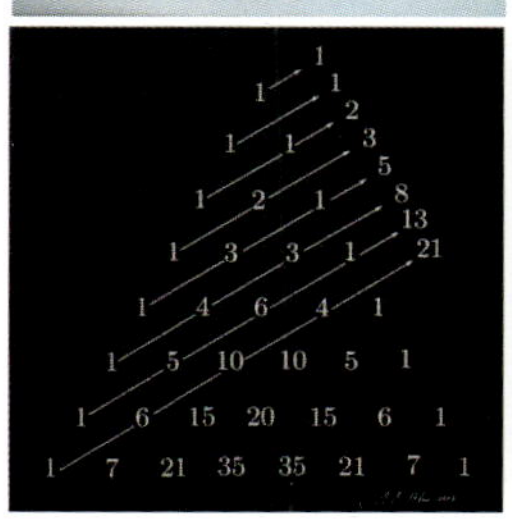

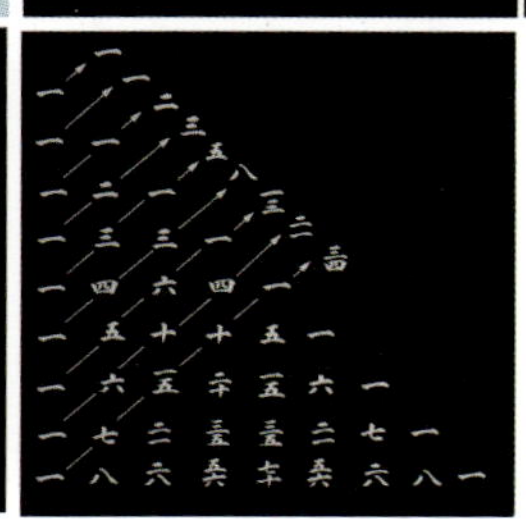

Renaissance artists inspired by Greek art and architecture use the golden mean proportions extensively. Plate 15 is an analysis of the famous altarpiece of Rogier van der Weyden's *Descent From the Cross.* One could also cite examples by Raphael, da Vinci and, more recently, Cezanne and early 20th-century artists. Pierro della Francesca was a master mathematician as well as artistic genius. His development and skill in the use of perspective can be studied effectively in his great *Flagellation of Christ* in Urbino, Italy. In Plate 5, his head studies from *De Prospectiva Pingendi* are shown. Measurement and close observation provided a balance for his mathematical principles.

Organization of numbers in patterns and rules of notation gradually led to arithmetic. Further transformation moved the concept of number from quantity to calculation in the development of algebra. Interwoven in the pursuit of knowledge are interesting diversions such as Pascal's triangle (Pl. 18) and its Chinese equivalent (Pl. 17). The mathematical study of probability in gaming revealed some new properties in the arithmetic triangle. Later mathematicians observed relationships with Fibonacci numbers, which are indicated on the diagonal.

Plate 3: CHOREOGRAPHY
Lucinda Childs with RDT
& Alvin Ailey dancers

Leonardo Pisano, known as Fibonacci, was born in 1175 in Pisa and occupies a unique position in the history of mathematics. In his early publication, *Liber Abaci* of 1202, he advocated the adoption of the Hindu-Arabic system of numeration. His studies with Muslim teachers and travels with merchants influenced his appreciation of their system of calculation. He did significant work in many areas, but is best known for the sequence that bears his name and for generating specific numbers for the golden proportion. The sequence has been extensively studied for the analysis of plant forms in nature.

Rennaissance pictorial organization (Pl.15) and the logarithmic spiral (Pl.16) are based on Fibonacci proportions. Beginning this study one thinks of da Vinci's *Modular,* all too frequently exploited in insurance literature. My own computer-generated figures, developed for an early project, were finally selected for Plate 1.

In architecture from Alberti and Brunelleschi to Le Corbusier, the golden mean persisted as a design basic for proportion and expression and continues to some degree today. In painting, anti-formulistic directions, surrealism and other fashions temporarily dominate. However a mathematical base, whether explicit or implied, underlies much of today's art.

Plate 6: OLYMPIC RACERS

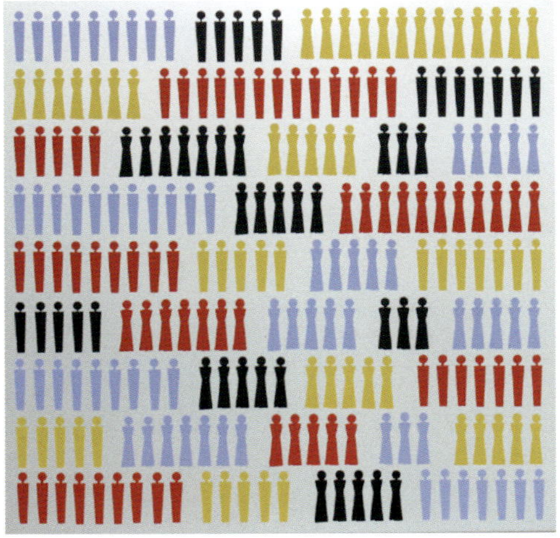

Plate 7: CENSUS

Plate 8: JAZZ by Dan Waldis

MEASURE

By expanding the concept of numbers to include measure, many related arts and cultures could be included. I asked the poet Katharine Coles, whose interests in science and mathematics were reflected in her work, if she would create a square poem about numbers. The resulting sonnet is a lively, engaging piece that easily could be the focus of a larger work. Dance was an area familiar to me and a surprise to many mathematicians. The choreography of Lucinda Childs is an excellent example of geometric patterning of movement. She graciously permitted me to include it here. Movement is also recorded as time in the swatch image of Olympic racers precisely caught by photography. Composer Dan Waldis was preparing a new work for the dedication of the Cowles Mathematics Building and contributed a few bars to represent music. All of these areas could be vastly expanded and perhaps a youthful mathematician will be inspired to do so.

NUMBERS

They won't stick. They gleam like brilliantine.
Perfect parsers, they jostle into essence
then reappear, renewed. A trillion seems
just so many zeroes. Xed-out, they dance,

Uncoupling and recoupling along a line
hoofing infinite movement, can-can's limber
leg and best foot forward, tapping time
until time is up and they're dismembered.

Dead-broke as syntax, clauses so declined
they tick themselves off. It's only beauty–
perfect measures measuring the mind–
mind tries to get around. Pen, brush, or flute.

Equation tooled to figure life. Amount
imagination multiplies. Takes to account.

Katharine Coles

NUMBERS/MEASURE/GEOMETRY

First Floor Corridor

Plates 19-27, 32-39

13th C. Yaqut Al-Mustasimi describes calligraphy as a spiritual geometry

Islamic Grill and Typical Pattern

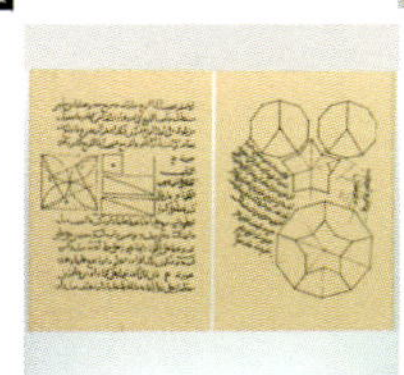

Early Islamic Manuscript

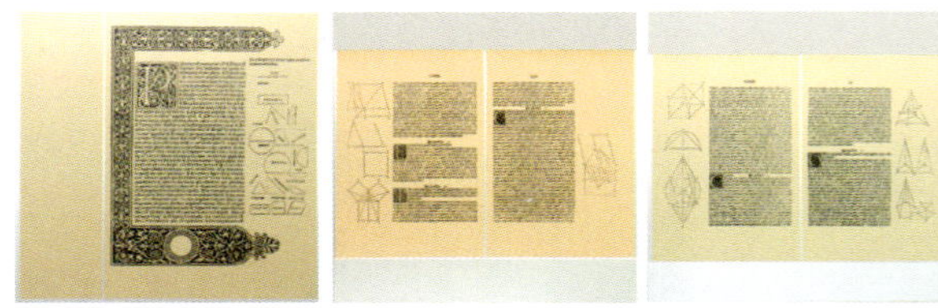

Euclid's Elementa Geometriae 1482

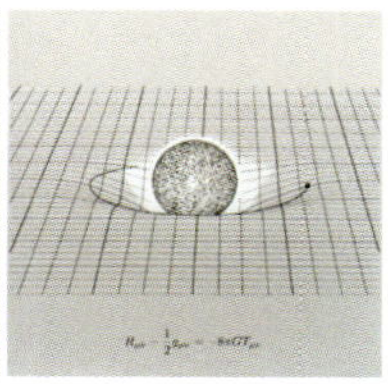

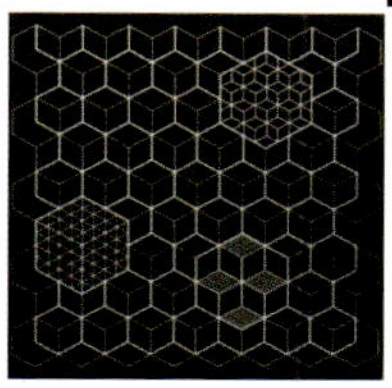

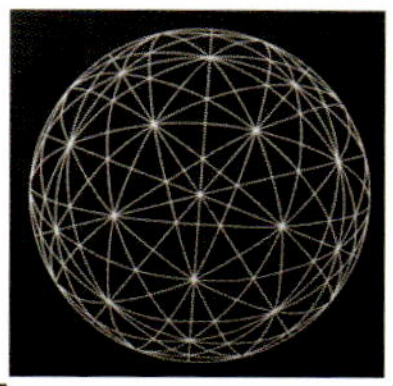

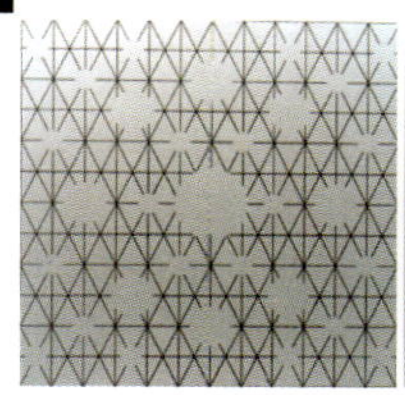

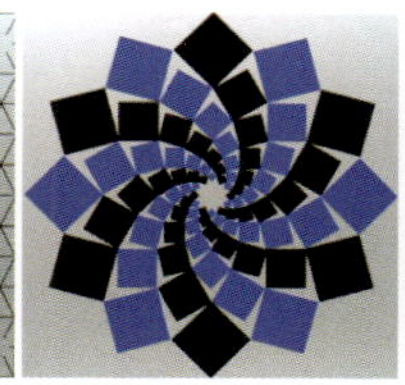

Moiré Study

Earth's Path with Einstein's Equation

Hexagonal Study

Spiral Geometry

Geodesic Dome

Implied Geomerty

Harmonic Growth

FASS Curve with figure ground reversal

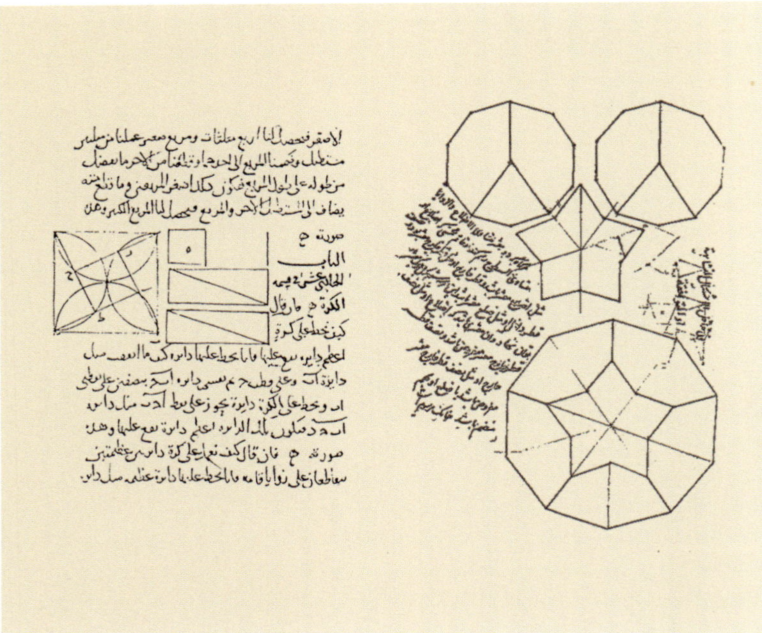

PLATE 24: ISLAMIC MANUSCRIPT

Calligraphy is beautiful writing and Islamic calligraphy is particularly so. As an introduction to Islam's influence in mathematics the text draws on the great 13th-century master Yaqui al-Musta simi: "Calligraphy is a spiritual geometry manifested by a material instrument" (Pl.19-21). While the spiritual side of mathematics is not addressed here, the work does not preclude it. The measured strokes and rhythmic beauty reveal more than words and complement the rigid geometry of Plates 22-23 and other panels.

PLATES 19-21: 13th C. YAQUT AL-MUSTASIMI DESCRIBES CALLIGRAPHY AS A SPIRITUAL GEOMETRY

Many of the early Islamic manuscripts I studied are devoted to Euclid's geometry. It seemed appropriate to include the original propositions in the edition of 1482 (Pl. 25-27). However the very early Paris Islamic geometric manuscript discovered by W. K. Chorbachi at the Biblioteque Nacionale also is important (Pl.24). Entitled *On the interlocking of similar congruent figures*, the drawing shows the design with the decagon and pentagon anticipating Penrose tiles.

Euclid's *Elementa Geometriae*, drawn from the Rare Books Collection of the J. Willard Marriott Library, is a treasure to enjoy in detail as well as to study the propositions. It is also one of the most significant books from the Gutenberg Press, published shortly after the Bible. Written in Latin, it is probably more accessible to students than a Greek edition would be.

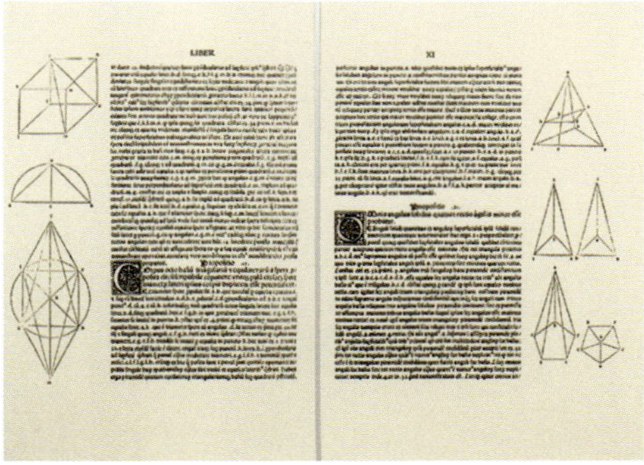

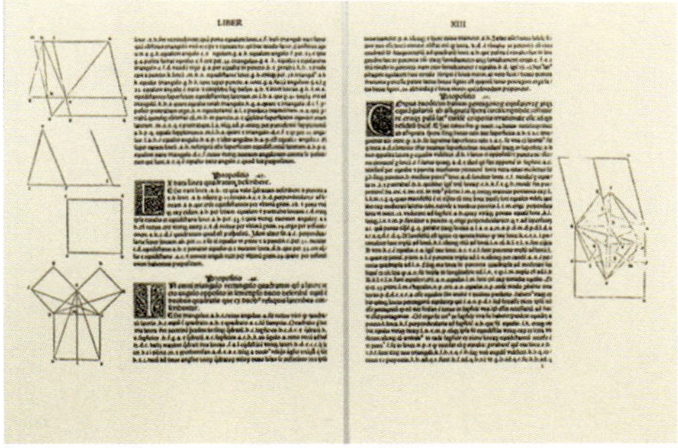

Plates 25-27: EUCLID Elementa Geometriae

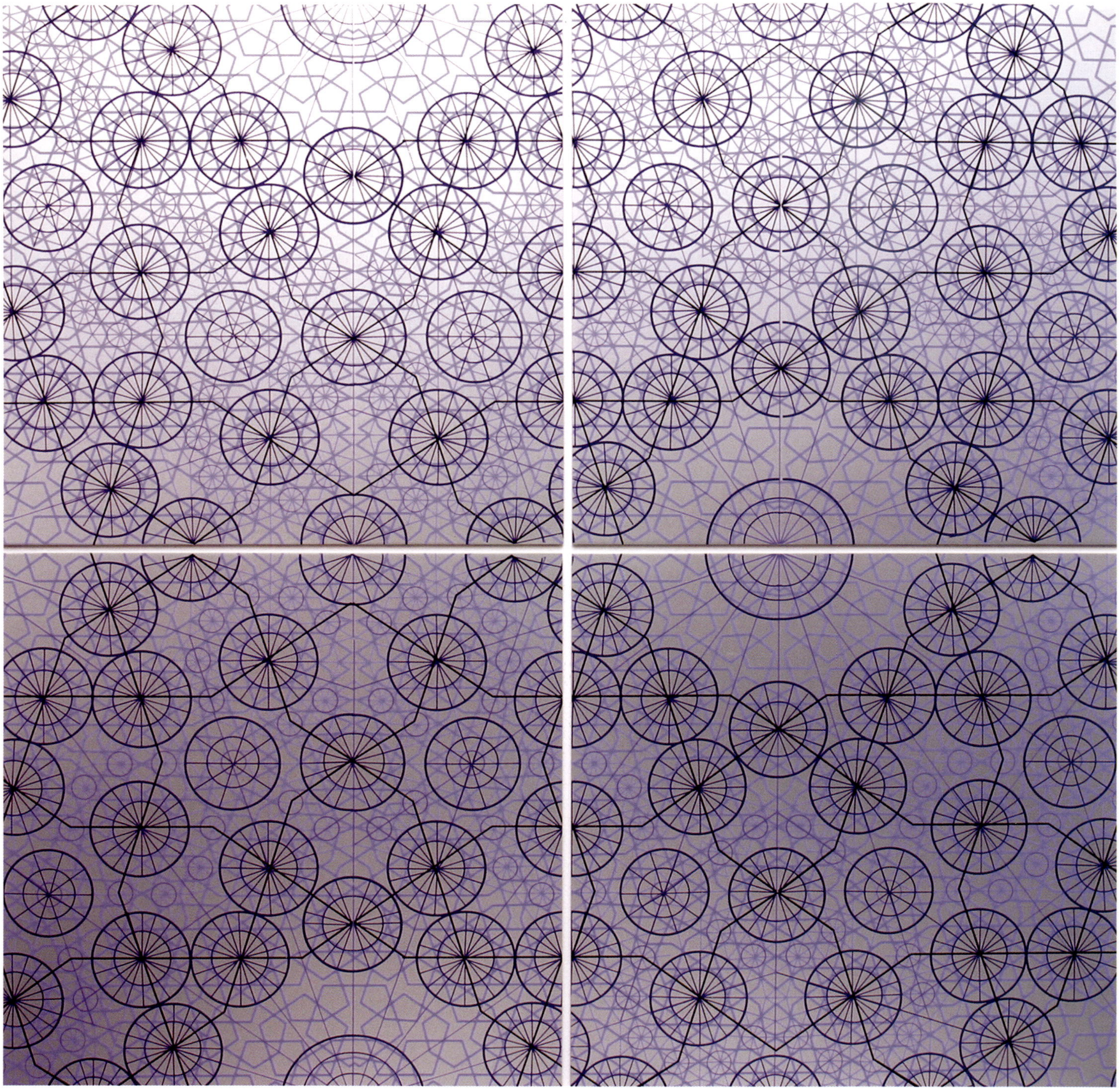

Plates 28-31:
TOPKAPI PALACE STUDY
Screen Printed on Aluminum

An excellent source for Islamic geometry is the Conference Proceedings of ISAMA-BRIDGES 2003."Meeting Alhambra" took place at the University of Granada, Spain, and at the site of the Alhambra.

Another magnificent book is *The Topkapi Scroll -- Geometry and Ornament in Islamic Architecture* by Gulru Necipoglu. The scroll's 114 patterns used at various times in the building of the Topkapi Palace are discussed, their importance for Islamic geometry and details of their construction. I drew inspiration from a small pattern expanded six times here (Pl.28-31). Separating layers for printing may be a more personal way of working than that of the original artist but the ultimate goal was to show the richness and beauty of the work.

Movement and spatial dimension in a a single plane also can be achieved by layering. In screen printing it is a natural process as each color occupies a separate layer. For the *TOPKAPI PALACE STUDY* (PL.28-31), seven intervals of subtle color changes were needed to give a fluctuating spatial quality. Achieving simple expression often involves complex actions.

Artists have long been fascinated with chance, movement, transformation and modularity. From Mozart to Cage and da Vinci to Johns, they have drawn on these concepts while creating their personal expression. In computer terminology one more likely would employ randomization, animation and other techniques to achieve similar concepts.

PLATE 22: ISLAMIC GRILL

PLATE 23: TYPICAL PATTERN

Plate 32: MOIRE' STUDY

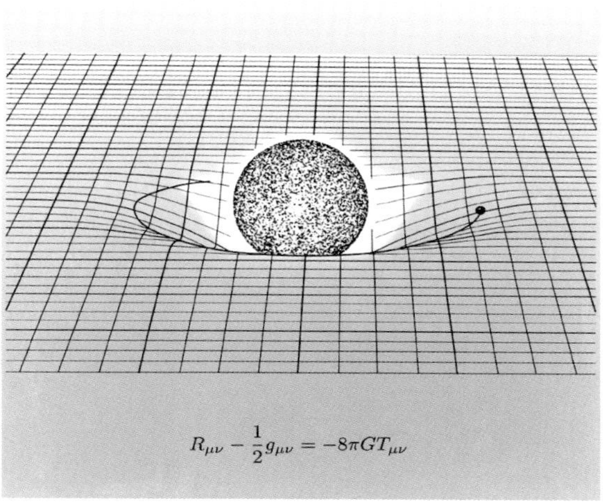

PLATE 33: EARTH'S PATH
WITH EINSTEIN'S EQUATION

PLATE 35: SPIRAL GEOMETRY

PLATE 36: GEODESIC DOME

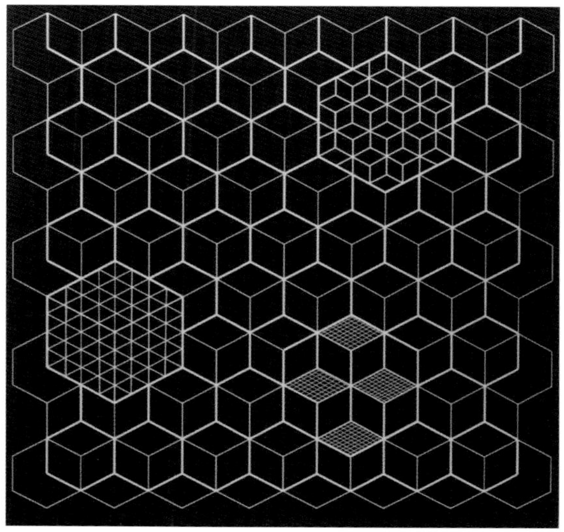

PLATE 34: HEXAGONAL STUDY

Plate 37: IMPLIED GEOMETRY

Geometry enters an artist's work on many different levels: at times as content, often as structure whether expressed or implied and occasionally as an intrusion. Including *IMPLIED GEOMETRY* (Pl. 37) in the corridor sequence can be challenged, as I know of no formula to explain it. The phenomenon is basically perceptual. Perhaps a future mathematician will define it.

During a series of studies using a triangular module and minimal lines, small circles of light appeared at the intersections. Instead of the layered veils of color I sought, the appearance was one of lace and I abandoned the prints. Later, in a large-scale painting, the same annoying flares occurred. The work was again based on a triangular module forming a hexagonal structure. The spokes were non-intersecting to prevent developing a fixed plane. They produced circular images of light, disturbing the spatial feeling I sought.

INTERSECTIONS WITH FLARE

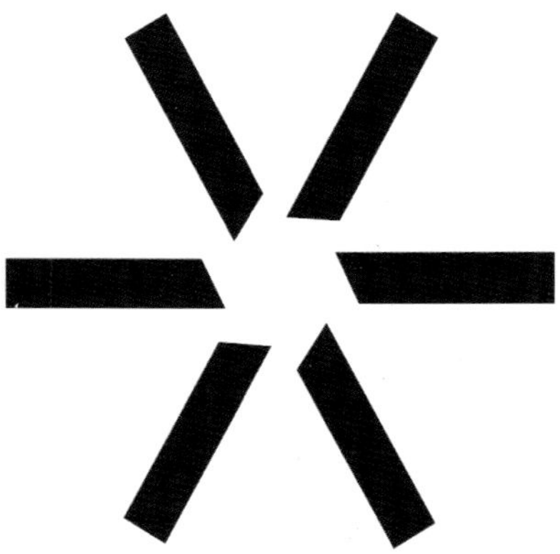

INTERSECTIONS WITHOUT FLARE

Several trials of the intersections suggested a way of drawing the ends at an angle, thus destroying the illusion. If you think of the circular flare as a bubble, then penetrating it destroys the illusion. Later I learned that the phenomenon is also known as subjective contours. Curiosity led me to examine it further.

ECHOES III on the following page shows the quality of space that can be achieved by modifying the grid intersections.

Working with grids and a modular structure comes from my architectural training and was important for this project in organizing a complex work. In another unit of the mural, I programmed a *FASS* drawing and converted it to a figure ground reversal by containing the ends and alternating black and white lines. Although fluctuating grounds are not new, having occurred in Greek, Islamic and Asian art, they are particularly strong having a geometric structure. Of particular interest to me are the illusions generating light, ambiguous spaces and sensation of movement. reinforced by geometry.

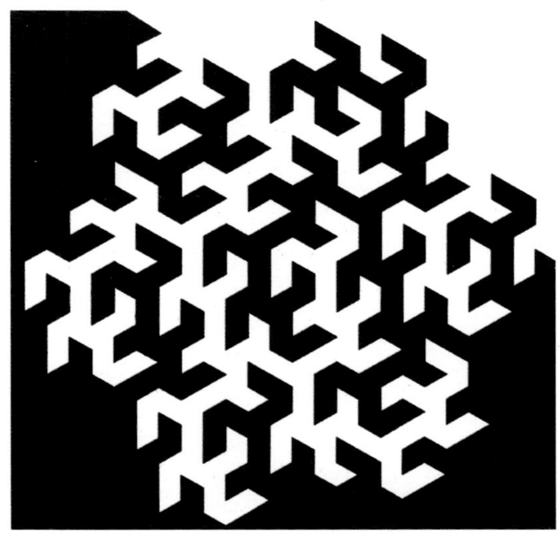

PLATE 39: FASS DRAWING WITH FIGURE GROUND REVERSAL

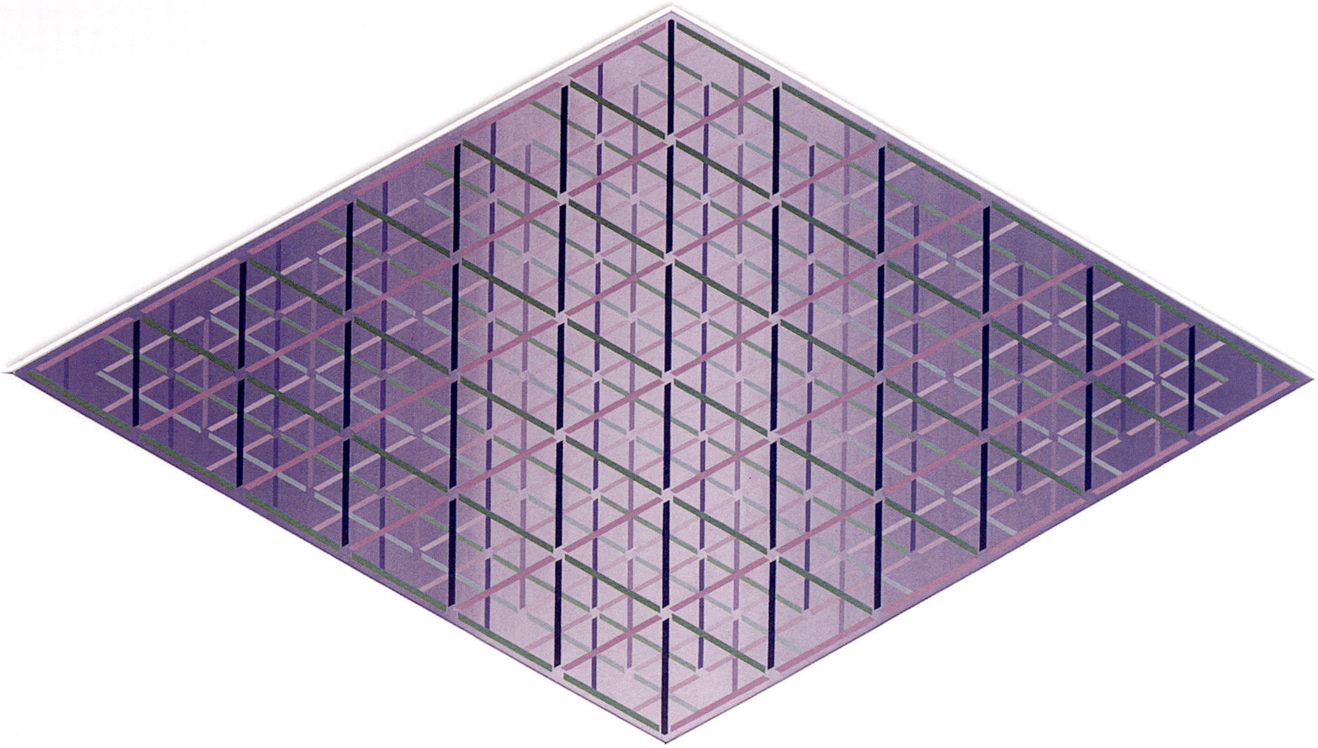

ECHOES III 1981-82, 40" x 70", oil on canvas

References for First Floor

1. Guedj, Denis, *NUMBERS: The Universal Language*. New York: Harry N. Abrams Inc. 1996.
2. Hampden-Turner, Charles, *Maps of the Mind*. New York: Collier Books, 1981.
3. Garland, Trudi Hammel, *Fascinating Fibonaccis.* Palo Alto: Seymour Publications, 1987.
4. Zakariya, Mohamed, *Music for the Eyes*. Los Angeles County Museum of Art Catalog, 1999.
5. Wade, David, *Patterns in Islamic Art.* New York: Overlook Press, 1976.
6. Hargittai, Istvan, *Symmetry 2.* Oxford: Pergamon Press, 1989.
7. Necipoglu, Gulru, *The Topkapi Scroll -- Geometry and Ornament in Islamic Architecture.* Santa Monica: The Getty Center for the History of Art and the Humanities, 1995.
8. Meeting Alhambra, ISAMA-BRIDGES 2003, *Conference Proceedings*, Javier Barrallo, Nathaniel Friedman et al, eds. (Granada, Spain: University of Granada, 2003).
9. Greene, Brian, *The Elegant Universe*. New York: W. W. Norton & Co., 1999.

PLATE 38: HARMONIC STUDY

INTERSECTIONS WITH NATURE

Second Floor Lobby

INTERSECTIONS is the theme for the second floor with the Lobby focus on nature. Flying over the country you are conscious of the circular patterns created by irrigation systems and the geometric divisions of property. Looking beyond our borders to maps of various countries, these divisions reveal much about the traditions of a culture, particularly the inheritance of land.

Nature offers a wide range of geometrical structures in plant forms explored here with fractal programming. The tree became the metaphor for many of the systems discussed in following pages.

SECOND FLOOR PLATES

1-4:	**Landscapes in Nature**
5-8:	**Water With Truchet Tiling**
9:	**Truchet Tile Study**
10-13:	**Fractal Plants and Trees**
14-15:	**Fractal Tree Branches**

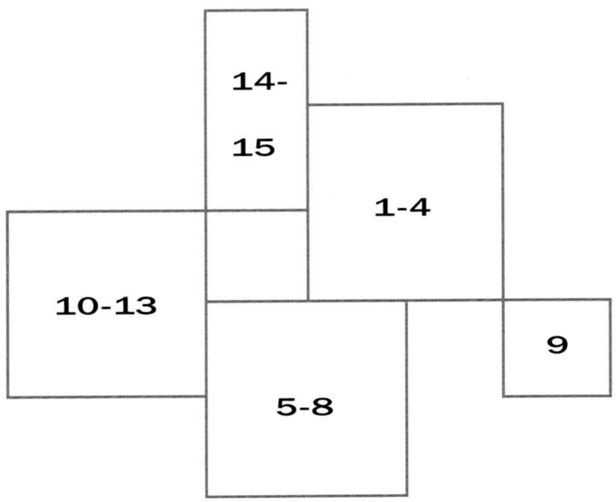

PLATES 5-8: TRUCHET WATER STUDY

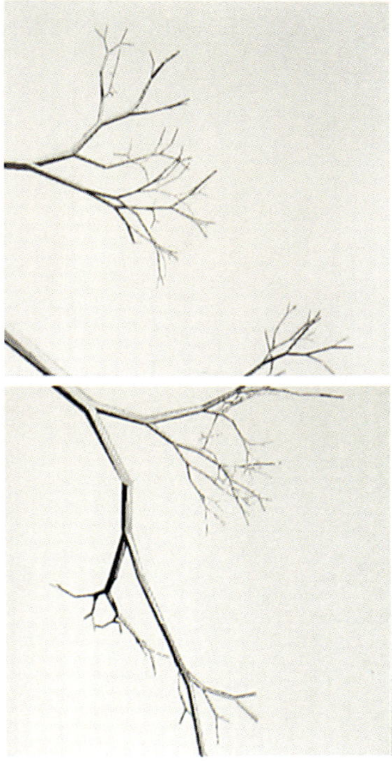

PLATES 14-15:

FRACTAL TREE BRANCHES

PLATE 9:

TRUCHET TILE STUDY

I think of the water study as Mythical Lake Truchet. It was inspired by the tile pattern created by the 18th-century French mathematician and memories of fishing on Puget Sound. Being an impatient fisherman, time between bites was spent observing the surface patterns of water moving in and out of focus. For this work the background was painted on the plates, the program generated on a computer and the image transferred to a screen for printing.

(Pl. 9) For the benefit of students, I showed the actual development of the tile pattern from two simple tiles with arcs in opposite corners. The random number generator in the program produced the pattern which to me suggests the movement of water.

We tend to organize visual material in a personal way. The introduction of random variations is a way of stepping out of a particular mind-set and often leads in interesting directions. Artists have used many devices for introducing chance elements in their work. Rolling dice may have been a means of achieving change for Mozart or Leonardo. Today, the computer offers numerous alternatives.

The ease with which an artist can use the computer to zoom in and out, enlarging or reducing images, is comparable to the way in which one views a work of art today.

Fractal geometry offers another way of approaching mathematics and exploring the visual world, where Euclidian geometry is limited in describing natural phenomena of some complexity, clouds, mountains, coastlines, etc. Fractals can create recursive mathematical models with self-similarity. Shapes with these characteristics appear similar in various scales while preserving their resolution.

Inspired by a beautiful photo of broccoli coral that appeared to be a fractal image, colleague Wayne Rossberg and I tried to program it. Nature defeated us, but left a challenge for the future. Tree forms proved to be more accommodating, as did ferns. Plant modeling for biological and graphic purposes has developed enormously in recent years. The tendency toward realism has been avoided here to reveal structural patterns.

The theory of fractals existed earlier and was described by Benoit Mandelbrot in 1970. Its evolution is largely due to advances in computer technology and more sophisticated software. The potential for art as well as aspects of science is enormous. Famous artists of the past have made use of the technology available to them and created works characteristic of their epochs.

INTERSECTIONS

Second Floor

Plate 16:

DARWIN'S NOTATIONS

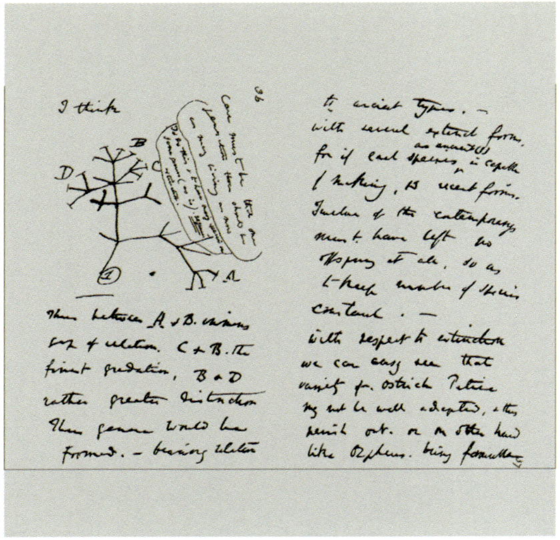

Several examples of rough tree sketches can be found in Darwin's notebooks showing his thinking while developing *On the Origin of Species.* Long the inspiration of poets and philosophers, it now can be considered a metaphor for scientific theory.

The elegant EVOLUTIONARY TREE of Sir Gavin de Beer, director of the British Museum (Pl.18), followed Darwin's theory. Linkages between various species were left unfixed due to lack of fossil evidence to support it.

The tree is also important for the concept of genetic mapping, currently an important area of research in many areas of Biology. Mapping for color inheritance in wheat is shown here in Plate 19.

The structure of DNA was verified by Watson and Crick in 1953 and led to experiments in genetic engineering. It is so well known that it is combined in Plate 20 with micro-arrays which suggest future individualized directions.

Plate 18: Left,1950

EVOLUTIONARY TREE
Sir Gavin de Beer

Plate 19: Right

GENETIC TREE

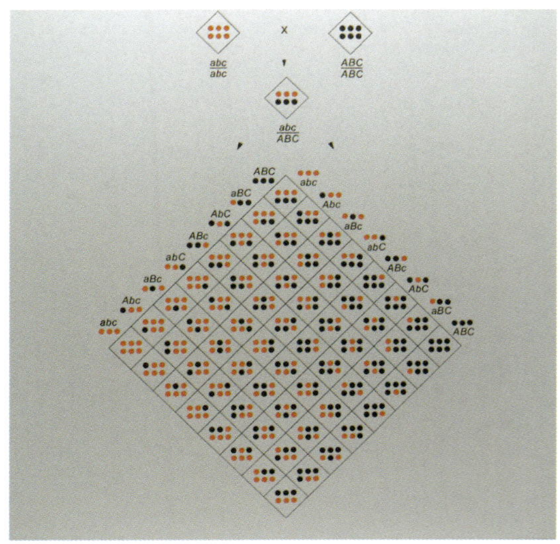

At first glance, Plates 22 and 17 raise an interesting question spanning both math and art. When does pattern become texture? Visually the surface of the Eye of the Drone Fly, though a small, regular pattern, inspires a sense of touch. Here it is magnified substantially. We don't think of mathematics as having sensory appeal although proportions can evoke certain feelings. Context and associations must be considered as contributing factors.

The drawing of the Eye of the Drone Fly is one of many beautiful studies from Robert Hooke's *Micrographia*. It was made possible by his invention of the compound microscope and engraved by outstanding artists whose pictures were included in the publication, a rare example of a scientist acknowledging his artists.

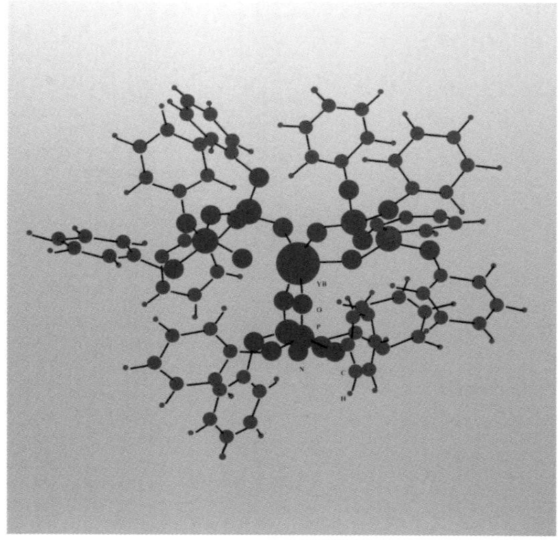

Plate 22:

MOLECULAR STRUCTURE of Silicon Carbide

The hexagonal pattern repeated irregularly in the silicon carbide molecule (Pl.22) can be described as a polytypic substance with order-disorder structures, consisting of layers periodic in two dimensions and stacked in different ways.

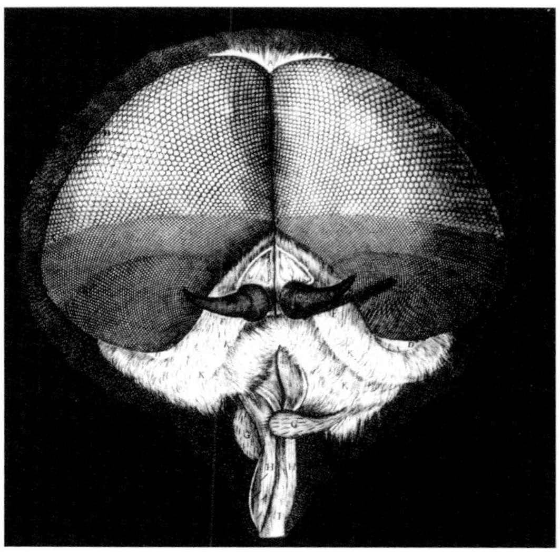

Plate 17: EYE OF THE DRONE FLY

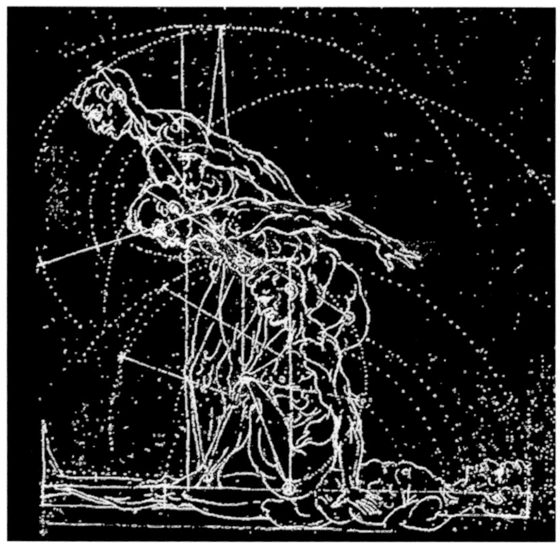

Plate 23: STUDY AFTER DA VINCI from the Codex Huygens

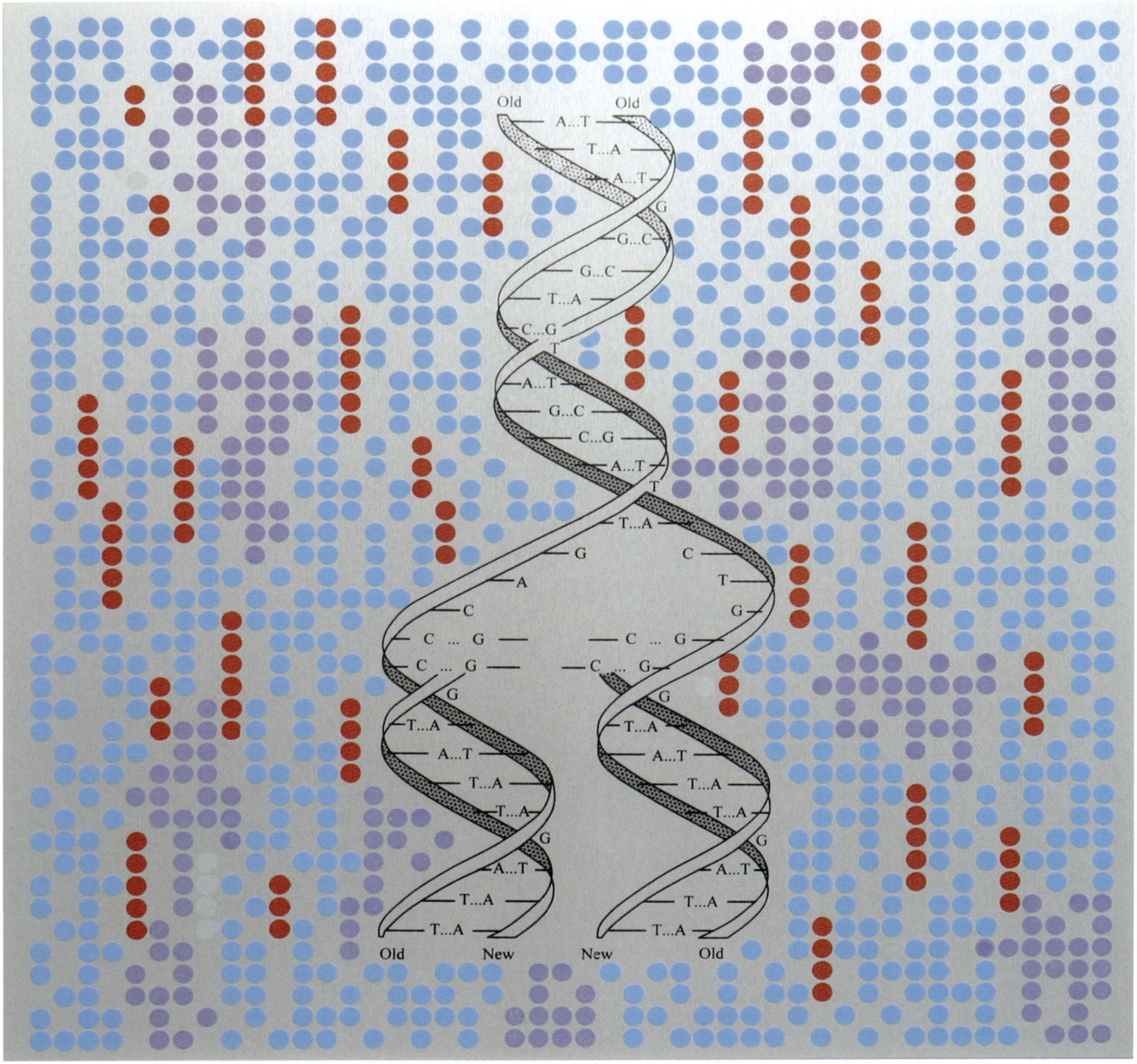

PLATE 20: DNA/MICROARRAY
SCREEN PRINTED ON ALUMINUM

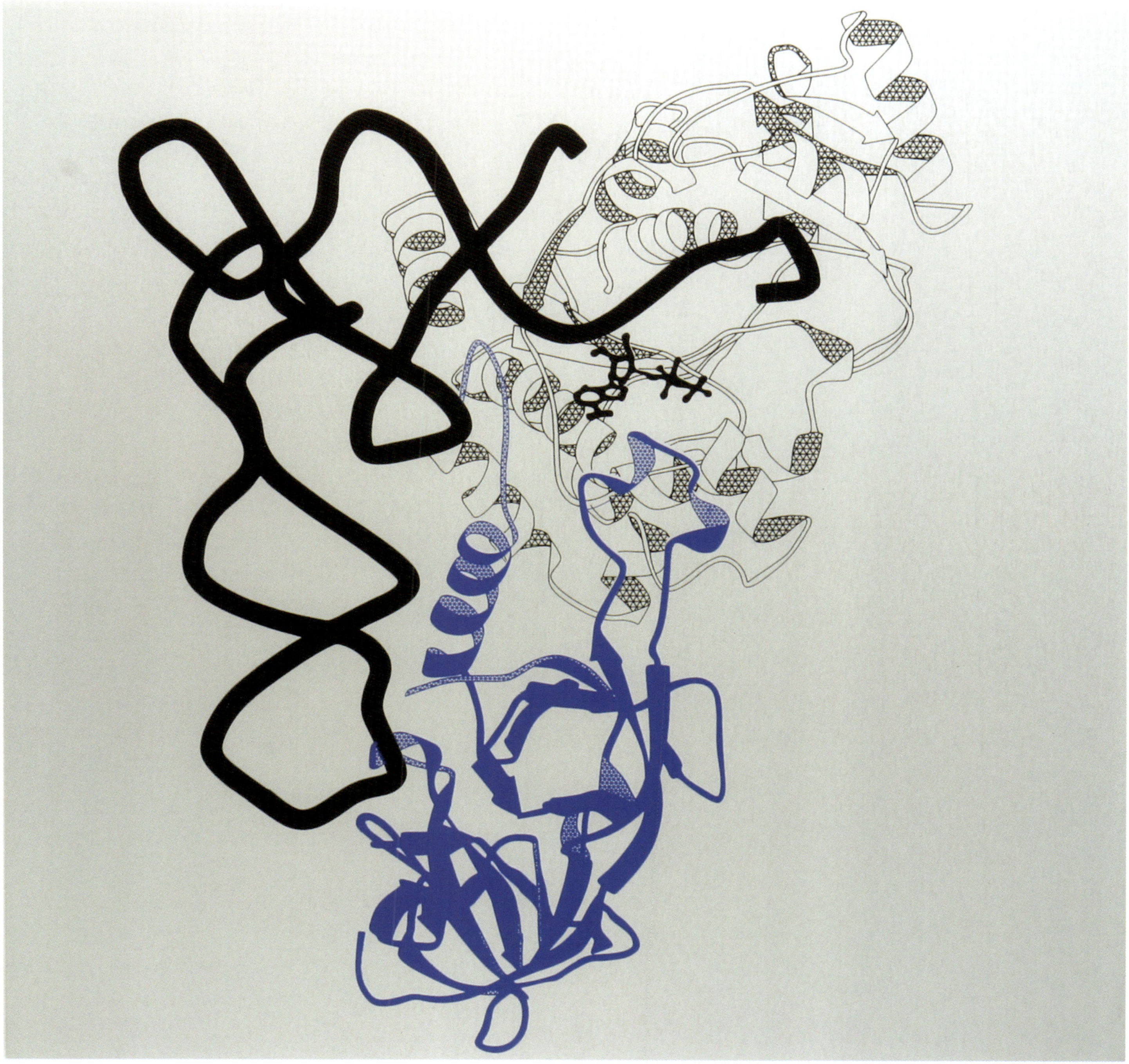

PLATE 21: RNA COMPLEX
SCREEN PRINTED ON ALUMINUM

*PLATE 24:

LIGHT WAVE INTERFERENCE

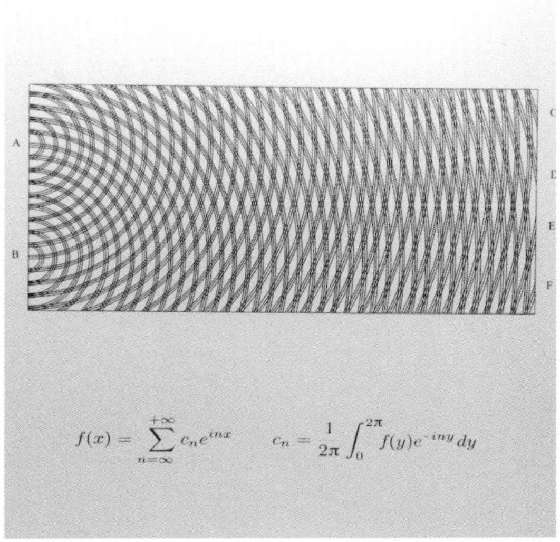

*PLATE 26:

ELECTRIC WAVES

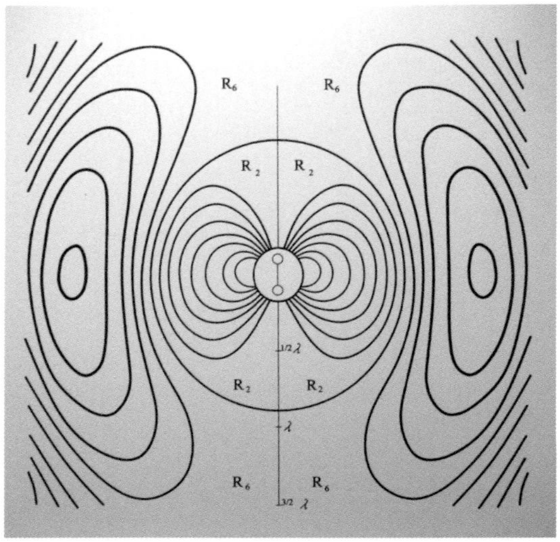

*PLATE 25:

ELECTROMAGNETIC FIELD

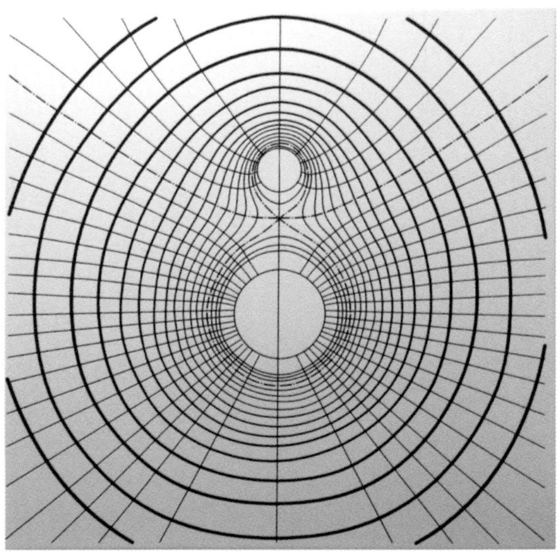

While mathematics is often the key to explaining phenomena, it is not easy to visualize without images. Thomas Young's diagram (Pl. 24) shows the analogy between light wave and water wave interference. Here two waves are examined oscillating at right angles in the direction of their propagation. The influence of Fourier's Transform Theorem as describing functions of time, temporal frequencies, and even spatial frequencies is suggested.

James Clark Maxwell in relating electric and magnetic phenomena described lines of force. (Pl. 25) Maxwell's Equations, a linked set of differential equations, relate the behavior of electric and magnetic fields and their interaction with matter. Heinrich Hertz further refined Maxwell's Equations (Pl.26) to take into account a different view of electromagnetism. The work of Maxwell and Hertz enabled Guglielmo Marchese Marconi, known as "father of the radio", to develop his system for wireless telegraphy.

The field of mathematics is so rich, varied and layered that one can not adequately trace its influence in the many areas of science. One can only suggest various instances in which advancements of science and math are inextricably linked. As new discoveries are made and theorems refined, the possibilities of further discovery are then broadened.

*Drawn from *The Scientific Image* by Harry Robin, 1992.

PLATES 28-29: SPECTRUM OF VISIBLE LIGHT, Young's Theory of B, G, R Receptors

Second Floor Science References

9. Prusinkiewicz, Przemyslaw, Aristid Lindenmayer, *The Algorithmic Beauty of Plants.* New York: Springler-Verlag, 1990.
10. Keeton & Gould, *Biological Science*. New York: W. W. Norton & Co., 1986.
11. Robin, Harry, *The Scientific Image from Cave to Computer.* New York: Abrams, 1992.
12. ISAMA - BRIDGES 2003; Meeting Alhambra; *Conference Proceedings*; University of Granada; Granada Spain
13. *The Magic of Microarrays.* Stephen H. Friend and Roland B. Stoughton, SCIENTIFIC AMERICAN, Feb. 2002, pp. 44-49.
14. Gesteland, Raymond F. and John F. Atkins, ed. *The RNA World.* Cold Spring Harbor: Laboratory Press, 1993.
15. Fichtner, Konrad, *Non-Space-Group Symmetry in Crystallography.* In *Symmetry,* edited by Istvan Hargittai, pp. 751-762. New York: Pergamon Press, 1986.
16. *The Tiling Patterns of Sebastian Truchet and the Topology of Structural Hierarchy*, C.S. Smith and Pauline Boucher, LEONARDO, vol. 20, no. 4, 1987, pp. 373-385.
17. Barnsley, Michael F., *Fractals Everywhere.* Boston: Academic Press, 1993.

PLATE 32: NIGERIAN TEXTILE DETAIL

INTERSECTIONS
with the Arts

PLATE 30: HAIDA DESIGN

PLATE 31: NAVAJO RUG DESIGN

This section of the mural posed a dilemma for me because the range of influence of mathematics in the arts is broad and deep although not always explicit. For my limited selections here I chose to emphasize so-called "primitive" arts because they illustrate the very basic geometry of vision. How we see and organize forms, how the structure of weaving has influenced the structure of buildings and, today, the developing computer system.

The art of the Haida of the Queen Charlotte Islands in the Pacific Northwest welds abstract forms with a sense of nature and a calligraphic hand. Plate 30 is from my own collection. The geometric forms of Navajo weaving (Pl. 31) have inspired many contemporary painters and weavers, just as African arts appealed to an earlier generation of cubists.

When we think of modular organization it is usually in terms of precise geometric forms. In ethnic arts, the boundaries are often softer or more fluid, but the sense of structure exists. It is the deviation from a rigid structure and more spontaneous patterning that makes African designs so appealing. Plate 32 of a resist-dyed textile from Nigeria suggests an artist with exceptional control and a wonderful sense of humor.

In early college studies in art history I found surprising attitudes about "primitive" art and later "mathematical" art. They were considered inferior forms or as irrelevant to the great history of art. As a result they became increasingly attractive to me. My experience of African, pre-Columbian and American Indian arts developed early, and Islamic art later through museum visits and travels in Spain and Turkey..

From weaving, dying, quilt-making and beadwork to wall painting (Pl. 33) is a natural progression for the women of the Ndebele people in South Africa. This has also been a means of preserving cultural identity through their numerous times of persecution, war and economic displacement.

Like so many of the African peoples, the Ndebele tribe holds a strong belief in the existence of a spirit world. Important in this world are the ancestors of the living who demand the continuation of tradition. While belief for many has waned, the observances continue. Women may paint their walls to please their ancestors and protect the house.

Today, strong color and geometric design characterize Ndebele wall painting. Working without benefit of preliminary drawings or rulers, the women paint directly on the walls. Earlier work made use of earth colors derived from local clays. Commercial paints now are generally used. Although more costly, they last longer.

Inspiration for their paintings may come from their highly developed beadwork, although the reverse may also occur. Today they are beginning to adopt motifs from their surrounding environments and one can expect this trend to continue.

PLATE 33: NDENBELE ENTRANCE

Native Habitats, (Pl. 34-35) seemed to be a natural link with the structures of Buckminster Fuller, Frank Lloyd Wright and the geometry of later architecture.

Traditional studies of architectural history generally start with the Egyptians and Greeks, giving emphasis to public and religious buildings and the houses of the rich. Ignored is the realm of the poor. Scant attention is given to prehistoric, Indian, African, Asian and nomadic groups. Although economically poor, their shelters are rich in concepts and sense of community.

Adapting to climatic needs early people constructed simple geometric forms for shelter from the elements, for protection from hostile groups and often on difficult terrain. Available materials varied with the region but ingenious solutions were found with reeds, clay and stone. Portability was another consideration for nomadic groups who had to move animals for seasonal grazing and find water for crops. Decoration on clay is often geometric and distinguishes a group's identity.

The construction of the Yurt shown in Pl. 34 reveals a sophisticated structure that is portable and adapts to many temperatures. The study by Hallet and Samizay, *Traditional Architecture of Afghanistan* details its use in various climates.

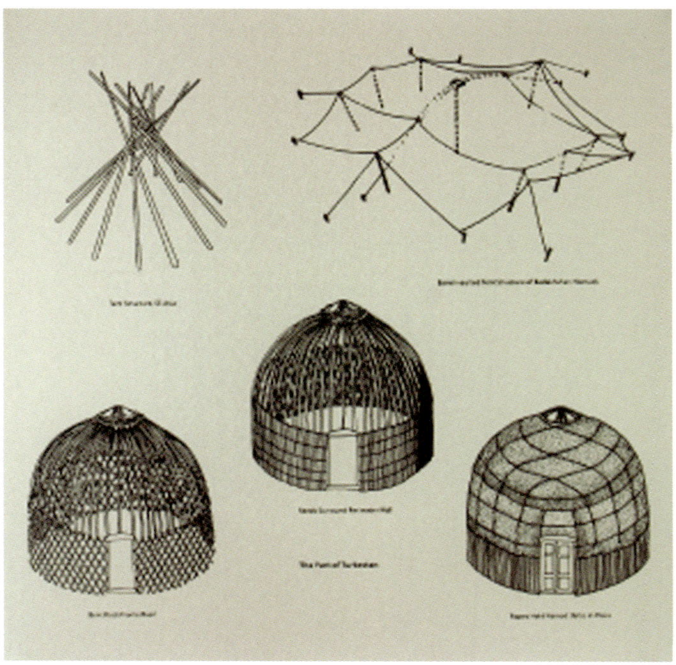

PLATES 34-35:

NATIVE HABITATS

42

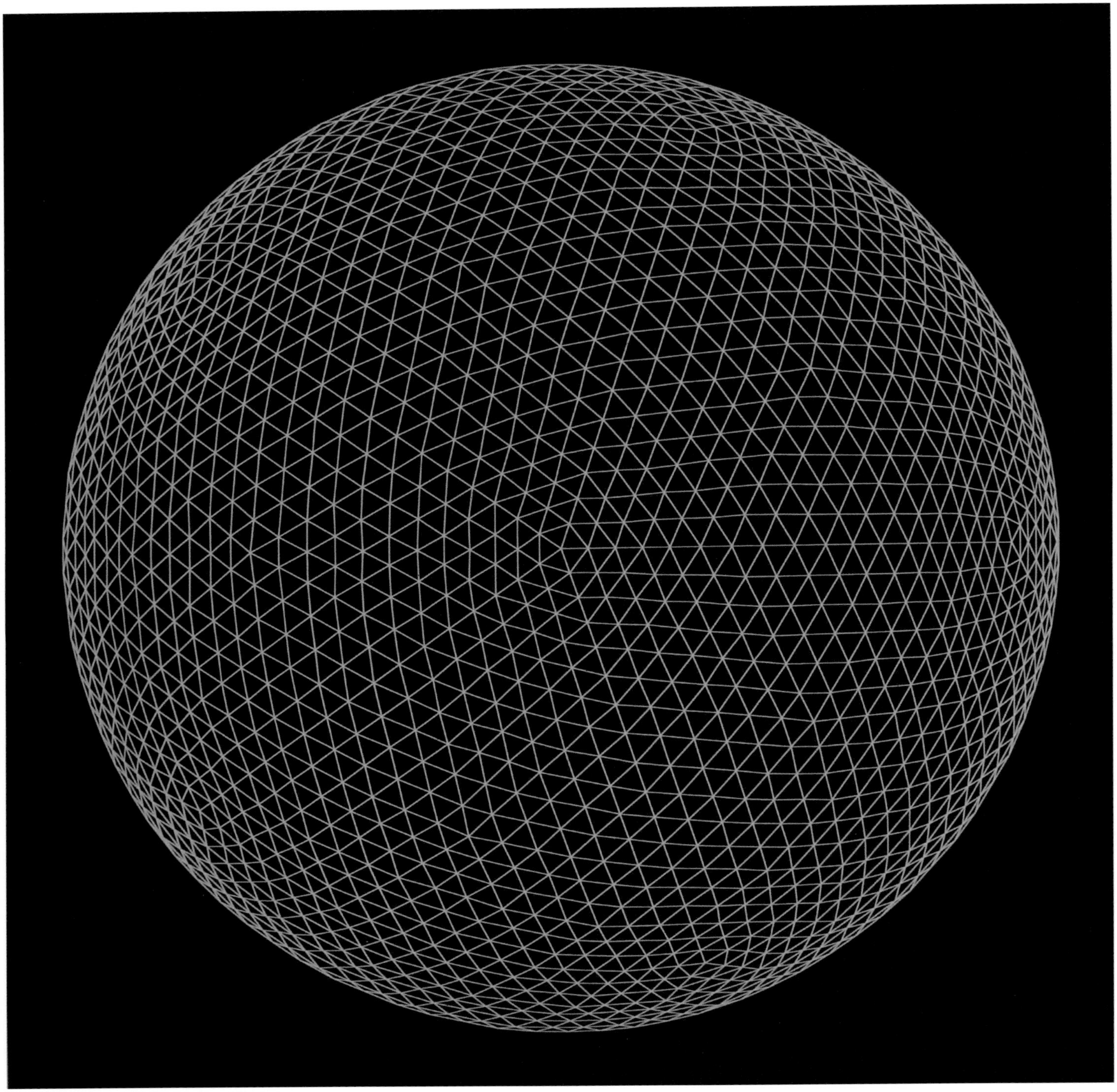

PLATE 36: R. BUCKMINSTER FULLER & SHOJI SADEO: U.S. PAVILION, MONTREAL 1967

To select a few masters to represent the range and influence of mathematics in architecture is naturally a subjective decision. What is important here is to make the connection and to suggest a range of geometry.

R. Buckminster Fuller's work reflects his intense observation of natural forms, his exploration of materials in model form and relentless pursuit of ideas about shelter. In some respects he shares a kinship with the builders of early habitats. Economy of means with available materials and geometric structures produce forms of great beauty.

Some of Fuller's concerns are shared with Frank Lloyd Wright, whose brilliance as a designer and engineer is part of history. Wright early focused on houses, their sense of shelter and relationship to the environment. His range of work moved well beyond domestic structures to urban concepts and planning, and using varied geometrical and structural themes.

Comparing the relation of the hexagonal module in the final plan for the Hanna House (Pl. 38) with the square module of Mies' famous pavilion in Barcelona (Pl. 39) provides some insight into the nature of architectural expression. The serenity achieved in Mies' work is also due to a conscious elimination of the unnecessary.

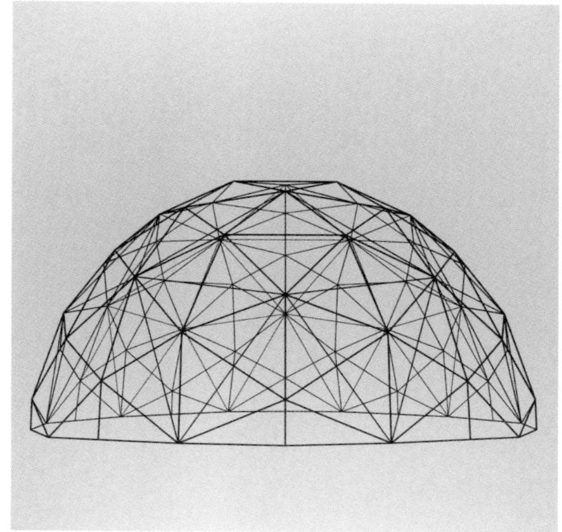

PLATE 37: Top

FULLER: GEODESIC DOME
1956

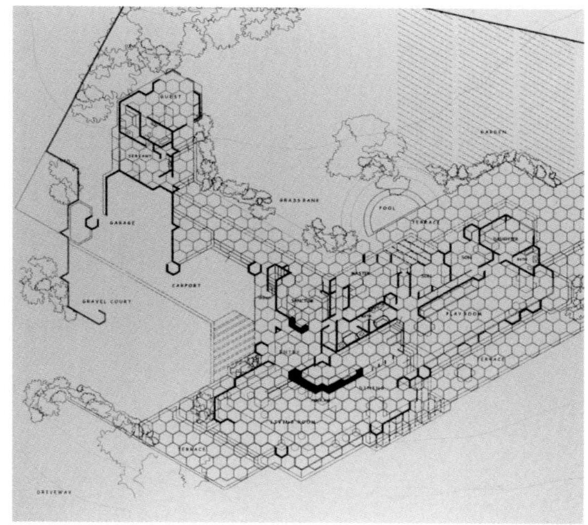

PLATE 38: Middle

FRANK LLOYD WRIGHT:
HANNA HOUSE, 1936

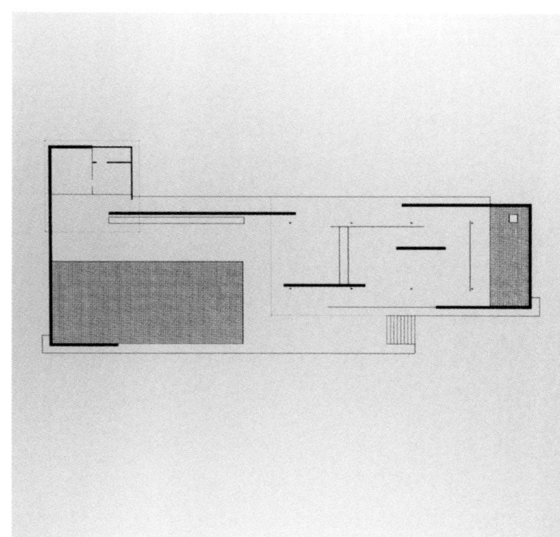

PLATE 39: Bottom

MIES VAN DER ROHE:
GERMAN PAVILION
BARCELONA, 1929

SANTIAGO CALITRAVA

One of the greatest living architects, Santiago Calatrava is a brilliant engineer who details his designs with the precision of a jeweler. The resulting structures rather than being sterile are elegant and daring while retaining a strong affinity for nature (Pl. 40).

Curator Matilda McQuaid has commented:

"Human endeavors are enriched by an intuitive vision that draws on all experienced and studied phenomena; such vision, as much as structural theory and geometry, can inspire monumental works of architecture. This belief underlies the work of Spanish architect and engineer Santiago Calatrava (b. 1951), whose expressive use of technology and inventive form reach beyond the realms of architecture and engineering. Music, painting and the natural sciences are as vital to his work as any calculation. The landscaped structures that he creates evoke an empathic response, reaffirming a place for awe in the criteria for building."

Second Floor Arts and Architecture

17. Washburn, Dorothy K. & Donald W. Crowe, *Symmetries of Culture,* Seattle: U. of Washington Press, 1988.
18. Picton, John & John Mack, *African Textiles.* London: British Museum Publishing, 1979.
19. Powell, Ivor, *Ndebele- A People and Their Art.* New York: Cross River Press. 1995.
20. Oliver, Paul, ed., *Encyclopedia of Vernacular Architecture, Vol. 1-3.* Cambridge: Cambridge Univ. Press, 1997.
21. Hallet, Stanley & Rafi Samizay, *Traditional Architecture of Afganistan*. New York: Garland STPM Press, 1980.
22. Ward, James, ed., *The Geodesic Revolution, Vol. 3 & 4.* New York: Garland Pub., 1985.
23. Hanna, Paul R. & Jean S., *Frank Lloyd Wright's Hanna House.* Cambridge: MIT Press, 1981.
24. Speyer, A. James & H. Fred Koeper, *Mies van der Rohe*. Chicago: The Art Institute of Chicago, 1968.
25. Zardini, Mirko, ed., *Santiago Calatrava – Secret Sketchbook.* New York: Monacelli Press, 1996.
26. Tischhauser, A.& S. vo Moos, *Calatrava Public Building.* Basel: Birkhauser Verlag, 1998.

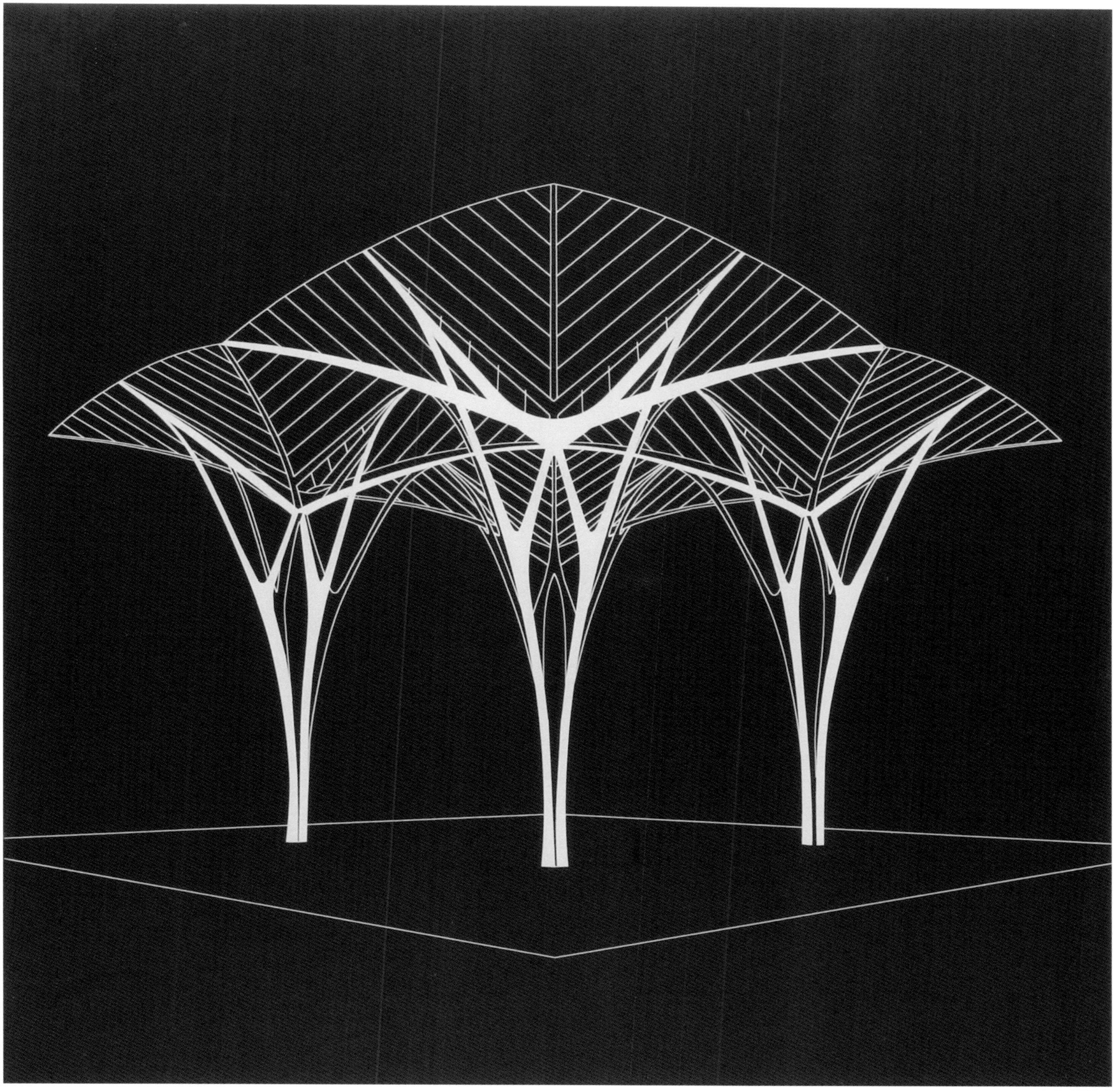

45

PLATE 40: SANTIAGO CALATRAVA: ZURICH PAVILION,1988
LASER ETCHING ON BLACK ANODIZED ALUMINUM

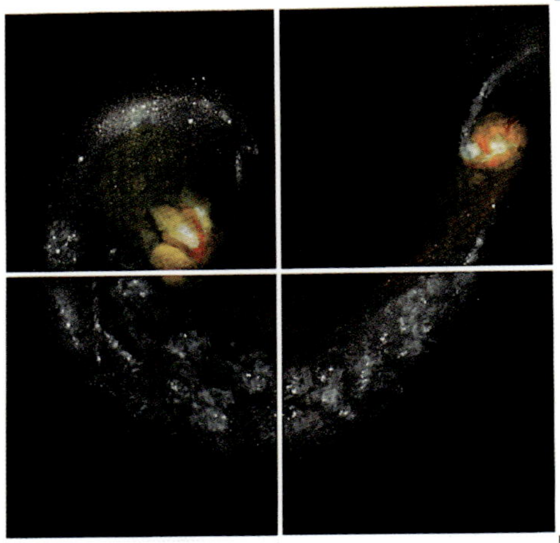

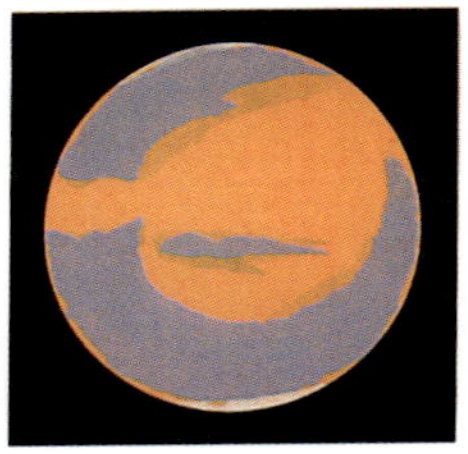

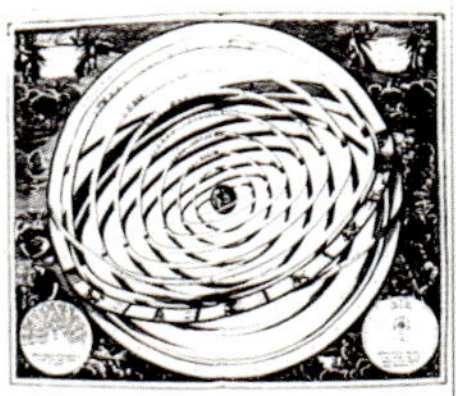

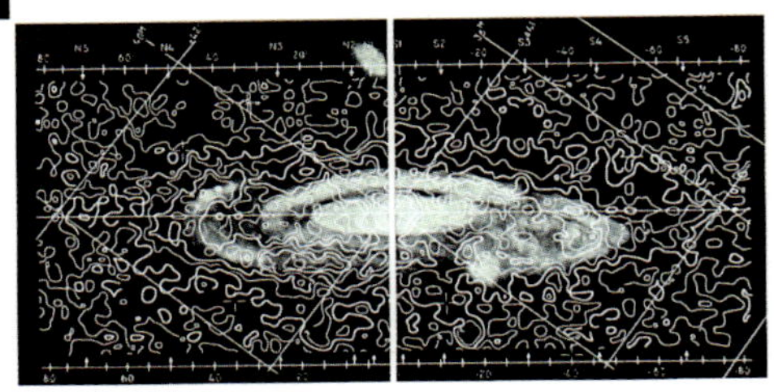

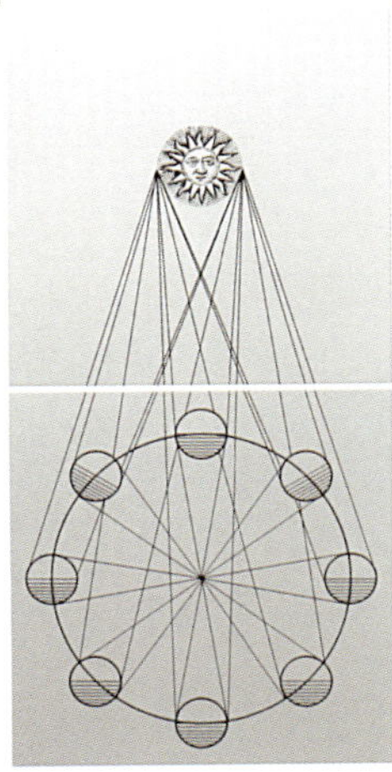

OUTER SPACE

Third Floor Lobby

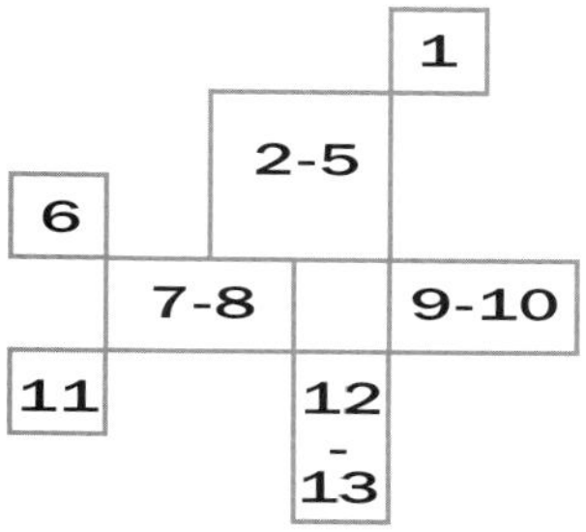

THIRD FLOOR PLATES

1: Constellation figures

2-5: Antennae Galaxy
from Hubble Telescope

6: Mars, July 2003

7: Ptolemaic system
of the world

8: Planetary orbit
Andreas Cellarius,1661

9-10: Radio map of
Andromeda Galaxy

11: Constellation figure
Sagittarius

12-13: Astronomical plan
Pierre Gassendi,1653

Mathematics has helped to raise the ceiling of our world from myth and theory to the experience of outer space. By adding precision to computer development and advanced telescopes, human exploration was made possible

We tend to think that our history began with the Greeks but we owe much to the early Egyptians, Mesopotamians, Indians and even earlier cultures. Prehistoric groups were concerned with cyclical patterns of weather for planting and harvesting while constellations helped with orientation and travel.

Prior to the invention of the telescope, information about the stars was limited by what could be seen with the unaided eye. With Galileo's observations and more advanced instruments, scientists' speculations could now be substantiated.

Medieval scholars regarded mathematics as the peak of human reason, surpassed only by faith. With the scientific advances of the 17th century, math became the language to express new observations and experiments in the physical sciences. Both Descartes and Newton used mathematics to describe concepts of space and movement in space.

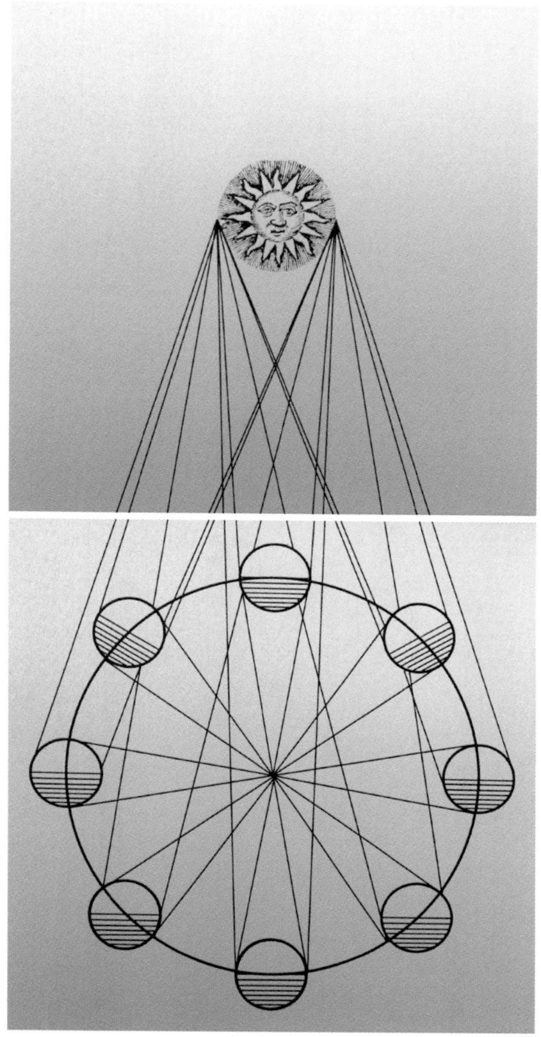

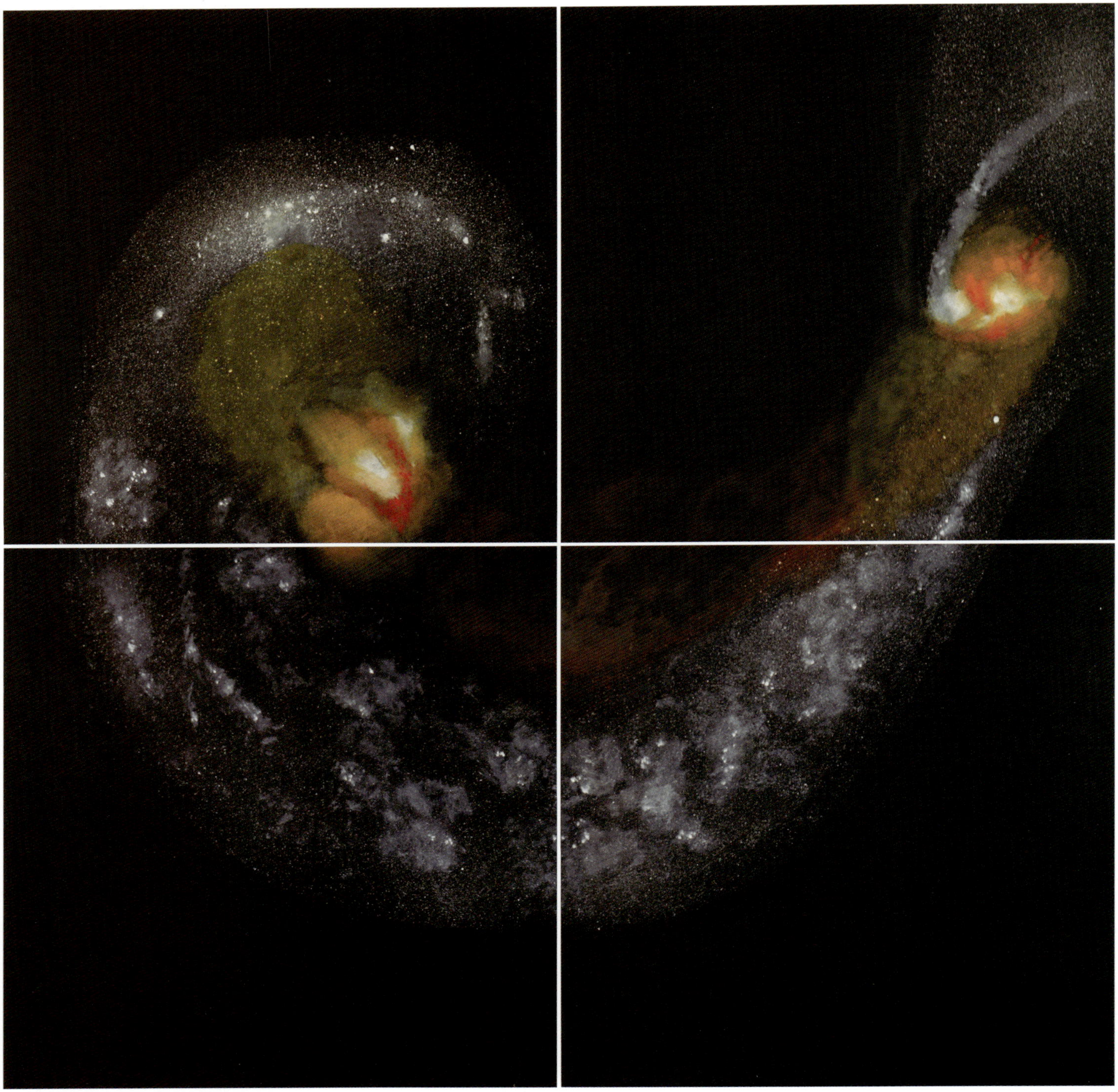

PLATES 2 - 5: ANTENNAE GALAXY from Hubble Telescope
OIL PAINT ON BLACK ANODIZED ALUMINUM

Scientists of the early 17th century had to confront three authorities: the Church, the Aristotelian orthodoxy at the Universities and the local astronomers working in a Ptolemic tradition. How they progressed in their work while satisfying the critics is an interesting historical study.

Petri Gassendi, whose astronomical plan of 1653 shown here (pl. 12-13), was a canon and later deacon of a church in Digne and professor of philosophy at the University of Aix until forced to leave by the Jesuit Order. His early published works were strongly anti-Scholastic. His later works focused on astronomy and issues of vision and motion.

Andreas Cellarius also bridged conflictng systems in *HARMONICA MACROCOSMICA* of 1661. Essentially a celestial atlas of Ptolemaic, Copernican, and Tychonic world systems. It illustrates advances in astronomical information and man's fascination with the universe.

In a giant step forward we come to the American astronomer, Edwin Powell Hubble who is noted for his discovery of galaxies beyond the Milky Way and establishing the foundation of today's contemporary physical cosmology. His design for the telescope has enabled NASA to uncover 1,000 bright, young star clusters. By probing the Antennae galaxies we have a new understanding of globular star clusters and with the beautiful photographs an appreciation of the beauty of the universe.

PLATE 7: PTOLEMAIC SYSTEM OF THE WORLD

PLATE 8: PLANETARY ORBIT ANDREAS CELLARIUS, 1661

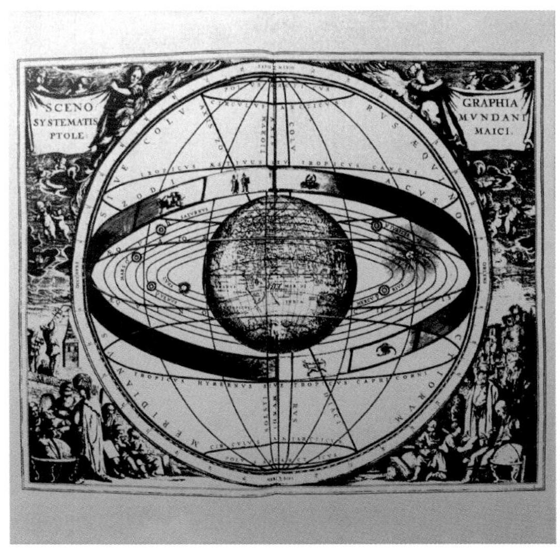

NAVIGATION BEFORE LONGITUDE

Third Floor Corridor

Plates 14 - 18

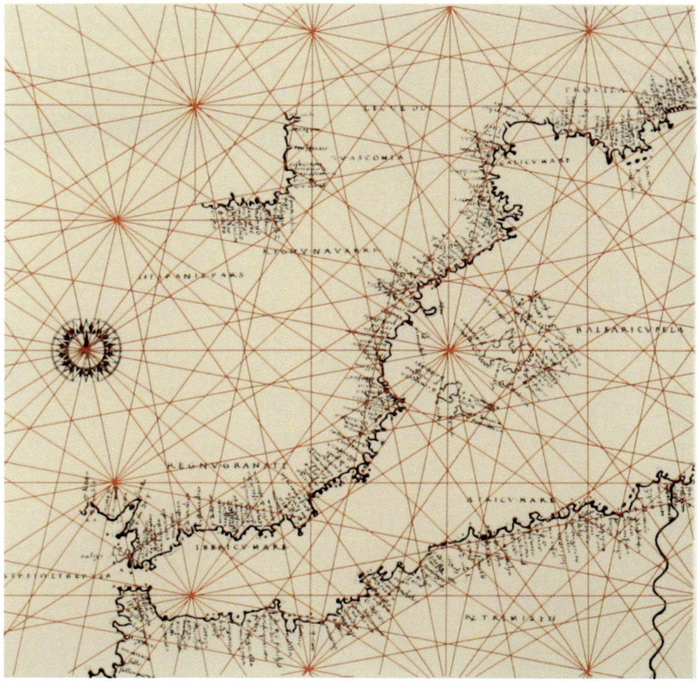

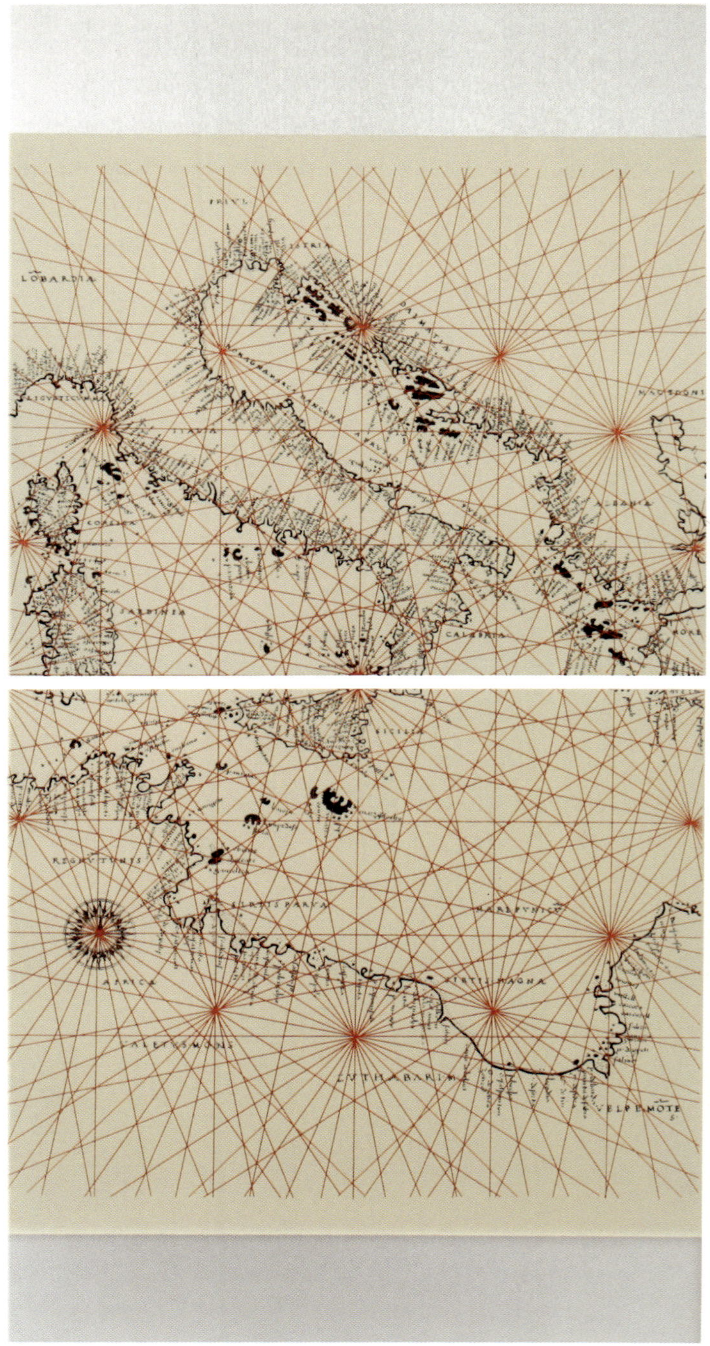

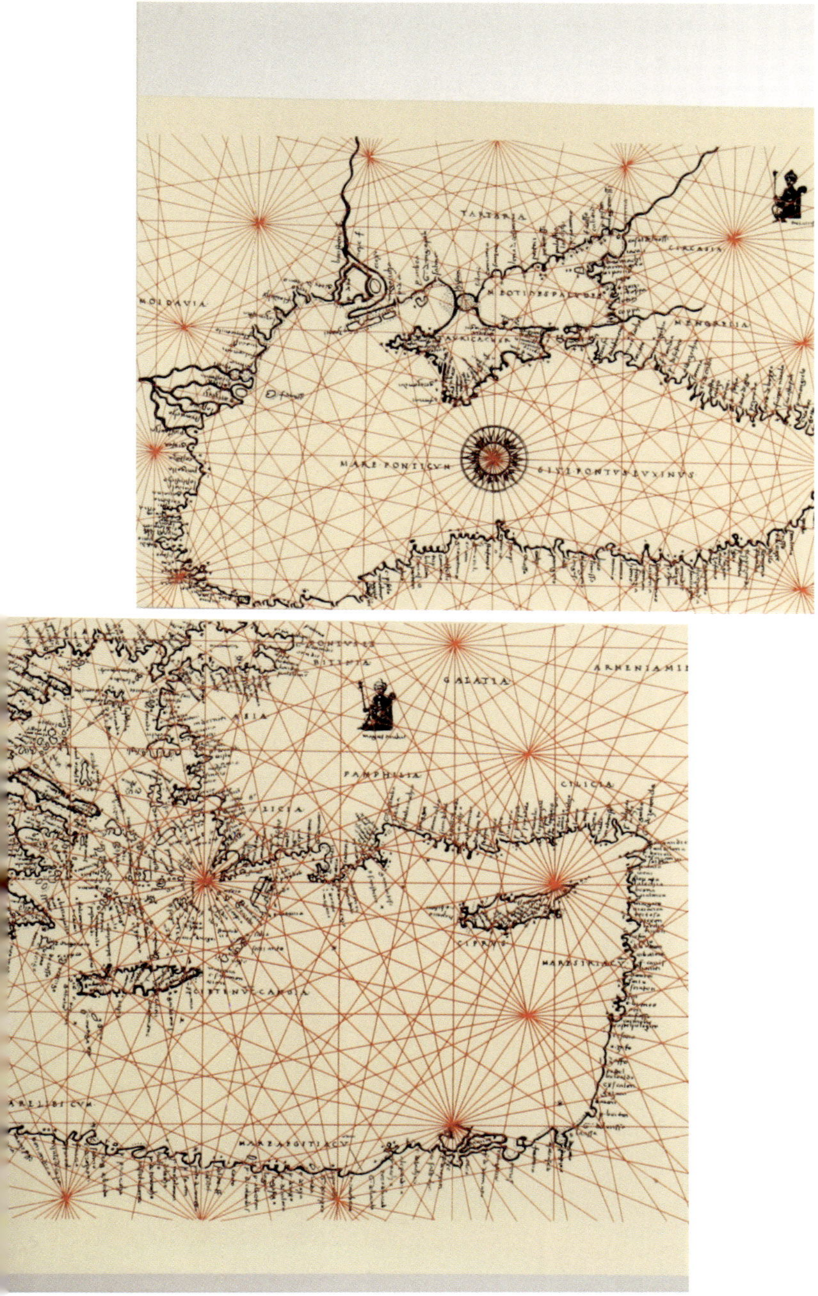

PORTOLAN-ATLAS, BATTISTA AGNESE 1546
COURTESY OF THE RARE BOOKS DIVISION
SPECIAL COLLECTIONS
j. WILLARD MARRIOT LIBRARY
screenprinted on aluminum plates

Portolan Charts provided sailing directions for pilots to go safely from one destination to another. They were valued for their accuracy as well as beauty and the quality of their design. Details of coastal features, distances, directions and harbors were provided. The magnetic compass was used at this time and points shown on the charts form a grid called rhumb lines.

Despite these aids, ships frequently were lost at sea. Storms and winds challenged the experience and judgment of mariners. Accurate instruments for determining longitude were lacking. It seems surprising that this was not solved until the 18th century.

On October 22, 1707, four British warships on the homeward journey ran aground . Nearly 2,000 men lost their lives. As a result, the search for a solution began to accelerate. The British Parliament in 1714 offered a prize (worth millions today) for a "Practicable and Useful" means of determining longitude.

The competition that followed is an absorbing tale of egos and foul play. John Harrison, a simple clockmaker of limited education but high intelligence produced the successful instrument. The scientific elite refused to acknowledge it, favoring astronomers over "mechanics." After 40 years, an aged Harrison finally received his monetary reward.

RHUMB LINES AS SHOWN IN PLATES 14-18

Third Floor References

27. Malin, David, *The Invisible Universe*. New York: IBS Books Stocked, 1999.

28. Agnese, Battista, *Der Portolan-Atlas des Battista Agnes von 1546*. Graz, Austria: Akademische Durck-u, Verlagsanstalt, 1993.

29. Sobel, Dava, *Longitude*. New York: Penguin Books USA Inc., 1996.

AFTERWORD

I have written these notes for students of all ages who have come to experience the mural. I've enjoyed the youngsters whose response was a simple "Wow." A group of 13-year-olds who came from Wyoming to see the mural for their current science study had much to say and to challenge. Older groups, museum volunteers, artists and visiting architects have all made discoveries, best expressed by a young faculty member: "I didn't know all of this was mathematics."

I approach art as exploration, owing much to training in art history, architecture and work in color and light at MIT. Painting, printing and computer studies followed. Intersections of Art/Science/Technology provide a broad base for my work. I seek connections between nature and the constructed environment, poetry and math. Use of the computer must be balanced with the more intuitive tradition of artistic expression. Often the most exciting ideas emerge at the intersection where more than one discipline meets.

In pursuing a technical field one needs depth and focus. I believe one should also develop wide-angle vision to observe the weeds at the edge of the field.

A. C. Bliss, 2006 *www.annacampbellbliss.com*

ACKNOWLEDGEMENTS

FOR the MURAL PROJECT:

Van Pilkington - anodized aluminum plates
Roy (Skip) Marsh - screen preparation
Aaron Engebretsen - laser etching
Michael E. Langton - photo stencils
Wayne Rossberg - computer programming
James Glenn - Utah Arts Council

STUDIO ASSISTANTS:

Skylar Nielsen
Edward S. Merrill
Tiffen West Roberts
Shana Kaufman

SUPPORT AND ADVICE for the BOOK:

Madelyn Garrett
Dorothy Greenland
Ted Nagata
Jill Schwartz
Reza Khazeni
Benjamin R. Butler
Elpitha D. Sifantonakis